P9-CBU-332

TEACHING MATH
to People with Down Syndrome and Other Hands-On Learners

Book 2
■ Advanced Survival Skills ■

DeAnna Horstmeier, Ph.D.

WOODBINE HOUSE ◆ 2008

© 2008 DeAnna Horstmeier
First edition

All rights reserved. Published in the United States of America by Woodbine House, Inc., 6510 Bells Mill Rd., Bethesda, MD 20817. 800-843-7323. www.woodbinehouse.com

The clipart used in Appendix B and on pages 16, 80, 83, 85, 99, 123, and 124 in this book and the accompanying CD are licensed from clipart.com.

Library of Congress Cataloging-in-Publication Data

Horstmeier, DeAnna
    Teaching math to people with Down syndrome and other hands-on learners / by DeAnna Horstmeier.—1st ed.
        v. cm.
    Includes bibliographical references and index.
ISBN 13: 978-1-890627-42-3 (bk. 1 : pbk.)
ISBN 10: 1-890627-42-9 (bk. 1 : pbk.)
1.   Mathematics—Study and teaching. 2. Down syndrome—Patients. I. Title.
    QA11.2.H67 2004
    510'.71—dc22   2004011436

ISBN Number for Volume 2: 978-1-890627-66-9

Manufactured in the United States of America

10 9 8 7 6 5 4 3 2 1

To the guys on North Street

Table of Contents

CHAPTER 1

Introduction to Teaching Advanced Math Survival Skills

Questions to be answered:

1. What is Survival Math?

2. Why are there two books on Teaching Math?

3. Why is there a need for these books?

4. Who will this book help?

5. What sets this book apart from other math books?

6. Who may teach the lessons in this book?

7. In what settings?

8. What areas of math are covered?

What Is Survival Math?

Teaching Math to People with Down Syndrome and Other Hands-On Learners, Book 1 introduced families and educators to the concept of *Survival Math*. Because hands-on learners, including those with Down syndrome, have only limited time to learn the many number skills that are necessary for a more independent life, the book concentrated on only those skills that were essential and focused on their application in daily life—thus the term *Survival Math*.

Why Are There Two Books on Teaching Math?

Even with a focus on math *survival* skills, it takes two books to teach the most essential concepts of math. Just think about how many textbooks are used in teaching math in general classrooms in both elementary and secondary schools—one book for every grade level, plus additional texts for geometry, algebra, and higher level math. In addition, now that more students with intellectual disabilities are being included in regular math classes at school, some of them need to learn certain skills and concepts to survive in the general education classroom.

The first book, *Teaching Math to People with Down Syndrome and Other Hands-On Learners, Book One,* was published in 2004 and covers number concepts, addition, subtraction, early money skills, time, measurement, and simple geometry. This second book covers multiplication and division, fractions, more advanced money skills, banking and shopping skills, and more information on measurement and time.

If you are not sure which book is appropriate for your student(s), both books include informal assessments to help you determine what skills a student has already mastered, or you may just skim over the table of contents or the first page of every chapter to get an idea of where to start.

Why Is There a Need for These Books?

As explained in Book 1 of *Teaching Math,* most students with Down syndrome, autism, or other cognitive disabilities do not learn or retain the math skills in school that they need to survive in the real world as adults. I have seen this firsthand with my young adult son, Scott, and also with other teenagers and adults. Although the current trend in education is to assure that all students get access to the general curriculum even if they have disabilities, this does not assure that they learn the functional math skills they will need to live and work independently as adults. Often students graduate from school not knowing how to:

- Use currency or coins in purchasing items;
- Use measuring cups in cooking;
- Write a check;
- Keep a record of what they are spending;
- Follow a simple budget;
- Add and subtract in simple daily living activities;
- Divide things equally.

Since 2000 I have been working with Scott and other young adults both in small groups and in a Young Adult Math class. In addition, I do professional development and college training with teachers of students with all ages. The teachers and I have been working on ways to help hands-on, concrete learners really understand useful math. One of the major ways that we have been able to convey useful knowledge to students who are having difficulties with math is by using concrete, hands-on materials (Heddens,1990). If the students can handle the materials and experience the actions, they are much more able to understand and use their math skills.

Parents, classroom teachers, and students have given me many ideas and have told me when I made mistakes. The *"survival skills"* approach in the *Teaching Math* books emphasizes:

- Hands-on-activities where the students touch and manipulate physical materials;
- A focus on the essential concepts, rather than just drill work with numbers;
- Games and other activities that enable the students to enjoy the additional practice that they need to make sure the concepts make it into long-term memory storage;
- Examples, hints, and teaching strategies that have already proved useful for other students;
- Motivation for the students by beginning lessons with success steps—activities that they can easily accomplish.

Who Will This Book Help?

Like the first book, this volume of *Teaching Math* is directed at students who learn through hands-on procedures, primarily students with Down syndrome, but also other groups of students who could benefit from these hands-on activities.

Much of the teaching of math concepts in school is done at the abstract level. Some individuals have difficulties with using abstract math skills in a meaningful way. The book's focus on the functional uses of math in daily living can be of use to various individuals, including:

- Students who are concrete thinkers
- Students with Down syndrome
- Some students with autism spectrum disorders

    (Although some students on the autism continuum may be able to memorize the math facts and perform calculations, they may have difficulty knowing how and when to use these skills.)
- Young adults with cognitive disabilities who are handicapped in their independent living by lack of math and budget skills
- Any elementary or secondary school students who have substantial difficulties with math computation and concepts
- Young children, who need deliberate, concrete instruction in early math concepts

    (When I conduct professional development workshops, teachers have told me that young children with and without cognitive delays find the concrete experiences and game format of this book a pleasant

way to learn. They are also progressing as concrete thinkers who can see concepts more easily when taught at their level.)

In writing this volume of *Teaching Math,* I became aware that the more advanced math skills taught in a functional, hands-on way were less geared to specific characteristics of students with Down syndrome than those in the first book. Just as the other books dealing with individuals with Down syndrome become more general when dealing with transition and adolescent issues, the strategies for learning for older students with Down syndrome are more general and apply to almost any student who learns in a hands-on method.

What Sets These Books Apart from Other Math Books?

- The early and frequent use of the **calculator**
- The emphasis on relevant, **functional skills**—for example, emphasis on quarters because of their frequent use in vending machines, and much less emphasis on pennies (Who really wants or needs pennies?)
- Emphasis on **problem solving** in **real situations** and not getting tied up in the language of word problems (In our own lives, how often do we have to solve a math problem from a written description in a book?)
- Activities designed to be **successful** with children who have difficulties learning abstract concepts
- Frequent use of **games** and **hands-on activities**
- An **informal assessment** that can be given as a series of **games**
- Activities appropriate for both **young and older learners**
- Use of **inexpensive, common materials** for the hands-on activities
- Simple, **clear instructions**
- Suggestions for ways these math concepts can be **woven into everyday living** (generalization)
- In Book 2, most of the skills are taught in **real-life situations** (for instance, fractions are taught by using measuring cups in baking)
- Also in Book 2, **a structured form** is used to help the students be able to use their computational skills in **real-life situations and story problems**

In *Teaching Math, Advanced Survival Skills,* concepts will be taught in two different ways:

1. by manipulating materials and helping the student to explore and understand a step-by-step process; and
2. by immersing the student in a real-life situation where he will actually see why this concept is needed.

Some students do well if we prepare them with hands-on materials where their focus is on that situation with little else getting their attention. After the student is secure with this one-on-one or small group procedure, we introduce him into a real-life situation with all its distractions and demands. Our preteaching of the concept will have made him feel more competent to deal with the real life situation.

Some students, however, can't see much use for learning the concept in a one-on-one or small group teaching situation, especially if they are older than elementary school age. Immersing them in a real situation might help them see that they need to learn more about the concept to be able to handle things they want to do in their real life. After participating in a real situation has motivated them, they may be more open to learning the skill or concept in a teaching situation.

Our ultimate goal, of course, is for students to understand what to do in various real-life settings. This specific type of generalization needs to be a major focus at this stage of teaching *Survival Math*. Whether you start with the teaching procedure or with the real-life situation will depend on what is most effective with your student or group of students. This book lists the teaching procedure first and then the real-life procedure, but you may arrange the procedures to fit your student(s). Games will also be used for a fun practice of the concepts. Other suggestions for generalization in daily activities will also be given. Parents will probably be able to put most activities into practice within their home activities.

Who May Teach the Lessons in These Books?

Any interested person can teach the lessons. Parents, educators, volunteers, siblings, or peers can do the activities with the student. The instructions should be clear enough for anyone to understand. However, I have discovered that "who" should teach these lessons may be different at different times in the student's life and according to very individual circumstances.

For example, some parents very much enjoyed doing the prenumber and early number lessons of Book 1 with their children as learning experiences and as games. However, when their children went to school, they preferred their children to learn at school and only used the lessons in the book when they felt their children were falling behind the rest of the class in learning math. Other parents have used the book for home schooling along with other workbooks, if necessary, for additional practice. Some parents have introduced the book to their children's teachers and both home and school worked together on hands-on activities, with parents doing follow through at home in daily activities.

Other teachers have done the hands-on lessons with their whole class and used the games for practice—and even as recreational activities when weather made outside recess impossible. Some teachers have used the book as a reference when some of the children were not getting a concept, and they thought that a hands-on experience might help. A few teachers have just handed the book to their classroom aide and told her to do the hands-on math activities with students who learn best with these kinds of activities. (Of course, I assume that the teacher supervises that teaching.) So it seems that the "who" that teaches the hands-on math varies with each individual situation.

In What Settings?

The survival math skills in this book are best taught with a group of two or three children. If the teaching is one-to-one, the instructor will have to play some of the games so the student has a competitor. Siblings or peer tutors may also participate. The activities in the book can definitely be taught to a larger group, but the instructor will not be as able to tell what the students' thought processes are. With a larger group, you should take time to informally evaluate each individual student's grasp of the concepts on a regular basis.

These books can be valuable for a student who receives math instruction in the general classroom. In other content areas such as science, the general classroom teacher can modify the goals of the instruction so that the student with special needs can participate and learn. In math, however, students with disabilities often need to master some of the earlier-taught concepts before they can understand the math activities of the general classroom. The activities in this book can be prepared and taught by a classroom aide, a volunteer, or an older peer tutor. The teacher just has to evaluate the student's progress and indicate the activities that should be done next. Some of the activities can be set up as a learning center for the entire classroom. Thus, this book can be a resource for appropriate learning for the child with special needs in the general classroom.

In a special education classroom, the groups are frequently smaller, and the lessons in this book can be adapted to the individual child's need. Care should be taken not to slow the pace of learning down due to expectations that are too low or because typical students are not present as models. More advanced students from older classes may be used to assist. The emphasis on relevant functional skills in both books is certainly helpful for students who are aiming for supported or independent living in the community after schooling.

Family one-to-one instruction can also be very effective. Students may be able to learn from the teacher and children in the regular classroom, but parents (or sometimes a tutor) can make sure that the major concepts are really understood. Teaching your child at home when he is young may enable him to participate more in the general classroom math instruction. Above all, families can make this instruction fun for the child and for themselves. Too often, both student and parent dread math homework. A special benefit can come from your family's knowledge of what motivates your child and of how to weave number activities into daily living.

Why Is the Subtitle of This Book *Advanced Survival Skills?*

Since I originally started working with four young adults who had been out of school for several years, I wanted to teach only those math skills that were necessary for their survival in the community. They had never mastered all the math facts, but they could use the calculator to substitute for this rote memory task. As I worked with them once or twice a week for over three years, I learned that they could be taught math concepts and skills. Why hadn't they learned them while they were in school? Or why had they forgotten them so soon?

Then my goal became to pass on some of the strategies for teaching math that had worked with these young adults to teachers of students of all ages. I wanted to make sure that future generations were taught in such a way that they could really understand math. I wanted to make sure that students with cognitive disabilities were taught the essentials of computation (using the calculator) while they were in school, as well as how to apply these skills to their daily living, especially in handling of money. Those are what I think of as "survival math" skills.

As I taught and investigated more, I realized that there were really two levels of survival math:

The first level, covered in Book 1 *(Basic Survival Skills),* encompasses skills that individuals need in order to handle the most common math-related tasks in daily life with the assistance of helping adults. Students at this level will:

- Have an understanding of what numbers and numerals are about.
- Be able to add and subtract.
- Be able to use these computation skills when they are needed.
- Experience some of the practical uses for math such as in measurement and in telling time.
- Understand the fundamental number concepts that structure our lives.

The student who has mastered these concepts can "survive" as an adult if he has someone to assist in setting up a structure for money handling. He will need family or providers to set up a budgeting system for him and guide him through a routine until he is secure. Family or other adults may need to do some oversight on his banking and his shopping experiences.

The second level, covered in this volume, *Advanced Survival Math,* progresses to using multiplication and simple division with the calculator for common situations and problems. More emphasis is given to teaching the student how to set up the problems that actually occur in his life. Many of the skills are taught in functional settings such as planning and buying several items independently and opening a checking account. Heavy emphasis is placed on money skills, along with the principles of budgeting, banking, and shopping. The student at this stage will be reasonably proficient at using the calculator but will need guidance and practice in applying his skills. Measurement, time, and fractions are explored in more depth. The student achieving Level 2 will be able use his math in routine situations in daily life at home, on the job, and in the community.

Have fun with this book! It is definitely *not* a workbook. Hopefully, you and your students will enjoy your learning together.

CHAPTER 2

Characteristics of Hands-On Learners

Questions to be answered:

1. What are the characteristics of most hands-on learners?

2. Why are some students concrete learners?

3. What are the learning characteristics of students with Down syndrome?
   - Short-term memory problems
   - Fine motor difficulties
   - Receptive language delays
   - Cognitive disabilities

4. What are the learning characteristics of students with autism?
   - Generalization difficulties
   - Receptive language understanding
   - Difficulties with imitation of others
   - Need for visual learning
   - Need for explicit teaching

5. What are the learning characteristic of students with learning disabilities?
   - Memory difficulties
   - Visual-spatial and organization problems
   - Difficulties focusing on learning

What Are Characteristics of Most Hands-on Learners?

In this book, the words *hands-on learners* are used to describe students who are concrete learners; that is, they learn best by actually handling and manipulating objects. In some cases, they may *only* be able to learn by handling objects. This is in contrast to many older children and adults who learn well by doing things concretely, but who are also able to picture and manipulate objects in their minds or use language to solve problems. Concrete learners can be learning to picture things visually and later to represent things symbolically, but essentially their knowledge is on a concrete, hands-on level.

Hands-on learners can be taught to manipulate abstract numerals, but they won't really understand the concept involved until they have direct experience with it. Some hands-on students can do pages of arithmetic problems, especially if they have access to a calculator, but have no idea how to add up the cost of several items they want to buy. Some students on the autism spectrum can easily memorize math facts and do rapid calculations but are not able to use math in daily life. They are hands-on learners as far as concepts are concerned.

Why Are Some Students Concrete Learners?

Jean Piaget, a famous Swiss psychologist, found that typical children go through developmental stages of thinking at particular ages. For instance, at the earliest stage (birth to 2 years for typical children), children learn through their senses and motor activity. At the "preoperational" level (2 to 7 years), children learn to imitate and to play with toys and others, and can use language to describe what they are doing. At about ages 7 to 11 years, children are in what is called the "concrete operational" stage. These children can think more logically than the pre-operational child, but they do have to relate their thoughts to visual images or objects.

At the concrete operational stage, students are far more effective and capable at problem solving than before. However, their problem solving suffers from one important limitation. They can solve problems in an organized way only when dealing with concrete information that they can see and perceive directly. Individuals in the concrete operational stage deal poorly with abstract ideas (Berk, 2001). For example, a child who is at the concrete operational stage may be able to look at a pie that has been divided into sixths and tell you how much of the pie is left when two pieces are taken away. But when confronted with a word problem that requires her to figure out that she needs to subtract 2/6 from 1, the problem may be meaningless to her. In other words, children who are pre-operational and concrete (operational) learners can compute with real things that they see, but have little chance of learning when given situations that require picturing the problem in their minds and doing mental math to solve it.

According to Piaget, from about 11 years on, most children are capable of thinking abstractly, reasoning with symbols, and thinking of most possible outcomes without constant reference to the real world about them.

Schools generally gear their curricula to the developmental ages that they expect typically developing children to have reached at any given grade level. Therefore, most elementary school age children are either preoperational or concrete operational level learners and learn best hands-on, at least initially. Thus, activities outlined in both *Teaching Math* books can be used for most elementary-aged school children.

Why Are Some Concrete Learners Reluctant to Do Math?

Sometimes older concrete learners may have had *so many experiences with frustration that they are not motivated to try* and give up easily or refuse to participate. They also may not see how the math activity could be important to them. These students are much more motivated if they can immediately see how that skill will help them in their daily life. We often call math that is rooted in real-life situations *Applied Math*.

Learning Characteristics of Students with Down Syndrome

Children with Down syndrome are mostly hands-on learners and also have other characteristics that need to be considered when teaching them about math. Research on the progress of students with Down syndrome in math is very limited. It is also quite probable that present-day adults with Down syndrome have had little systematic instruction in math and thus have not progressed up to their potential. Gillian Bird and Sue Buckley (educators and researchers at the Down Syndrome Educational Trust, UK, who have made it their life's work to help children with DS learn) have stated that in their clinical experience, "There is a wide variation in number ability among individuals with Down syndrome" (2001). This variation makes it very difficult to estimate what an individual person with Down syndrome is capable of doing in the math area.

In the absence of research pinpointing exactly why math is difficult for individuals with Down syndrome, this section offers some of the most common theories. Not all people with Down syndrome have all of these challenges to the same degree, and it is important for anyone working with a particular student to find out exactly which of these problems are stumbling blocks for that individual. In general, however, many children with Down syndrome have at least some of these challenges, and some children may have all of them.

SHORT-TERM AND WORKING MEMORY

Students with Down syndrome frequently have ***problems with short-term and working memory***. Short-term memory helps us to keep specific facts or items that we have just heard or observed in mind as we decide what to do with them. For example, you keep a telephone number in memory for a short time until you have used it or have decided to memorize it. Working memory usually contains general processes that you must remember as you work with short-term items. For example, a student needs to keep the steps in division in mind as she plugs in the specific data from the problem she has been given. In math, we often use short-term memory in learning numerals, computation facts, and specific details about the current problem that is to be solved. Working memory underpins the processes of math—addition, subtraction, multiplication, division, and other multi-step processes. Students with Down syndrome need assistance in learning to overcome these possible deficits in memory.

Of course, if students have difficulties with short-term and working memory, they will not be adept at putting facts into long-term memory storage. When information (such as math facts) is not of interest to them or has little emotional impact, they may seem to remember it one day and forget it the next. This is not willful forgetting on the child's part—instead, it just means that the information was never properly stored in her memory to begin with. However, when items are securely in long-term memory, children with Down syndrome can remember them for a long time—sometimes even longer than their parents do.

FINE MOTOR DIFFICULTIES

Many children with Down syndrome also have ***fine motor delays in their hands and problems with eye-hand coordination***. These delays make it difficult to manipulate objects and make writing numerals difficult and slow. My son says, "I just have fat fingers. I want them to be fast fingers." It may be that they do not get as much experience with exploring objects in their world if they have difficulty manipulating them. Certainly, they will have less energy to devote to understanding number relationships if they are struggling with just writing and lining up the numbers in a math problem.

RECEPTIVE LANGUAGE UNDERSTANDING

Processing information given orally can sometimes present another problem. Studies indicate that about 40 percent of individuals with Down syndrome have a mild hearing loss and 10 to 15 percent have a more severe hearing loss (Fowler, 1995). Certainly, if children can't hear verbal instructions, they will have difficulties understanding words and concepts and distinguishing between similar sounding words, such as "forty" and "fourteen." However, hearing loss does not tell the whole story, since even individuals with Down syndrome who do not have hearing losses usually have receptive language problems.

Although children with Down syndrome can understand much of what they hear, they seem to process speech slowly and tend to miss details and information that is given sequentially, as in directions. Most children with Down syndrome have enough of a delay in auditory processing of speech that they learn much better through their visual senses or through a combination of the various senses such as tactile (touch) and kinesthetic (movement). However, this delay in processing receptive language may not be apparent to parents or teachers because children with Down syndrome are very adept at reading body language and tone of voice.

Teachers who teach primarily by talking to the class can be very frustrating to students with Down syndrome. A student may be unable to keep up with processing the speech that she hears and may miss the important points the teacher is trying to make. If the teacher would just show her *real* dollars and cents rather than numbers on a page, she could make a picture in her mind. Sometimes she gets overwhelmed by the flood of speech and just tunes out. She may think, "Someone else in the class will just have to show me how to add those dollars and cents later."

The concept of concrete learners helps to explain why people with Down syndrome do much better when using manipulatives in math and are motivated by real-life situations even when they are in middle or high school. The real problem solving goes on with concrete materials that they can see.

Effect of Cognitive Disabilities

Cognitive disabilities have an undeniable effect on math learning. The ability to see the overall procedure, to discover patterns, and to generalize skills to other settings is definitely affected by difficulties with cognitive functioning. In addition, if a student's cognitive skills do not progress beyond the concrete operational stage identified by Piaget, she will struggle with the abstract reasoning needed for upper level math. Cognitive disabilities are usually a factor for students with DS and may be a factor for students with autism spectrum disorders. In addition, there are other students who have cognitive disabilities due to other or unknown causes.

Regardless of the cause of cognitive disabilities, they will cause difficulties with abstract concepts and reasoning, as well as some problems with attention, generalization, memory, and problem solving. For all these reasons, hands-on learning is most effective for students with cognitive disabilities.

Learning Characteristics of Students with Autism Spectrum Disorders

GENERALIZATION DIFFICULTIES

As mentioned above, students with autism spectrum disorders (ASD) may have difficulty applying their math skills to real-life situations. They frequently have difficulty with generalizing from one setting to another or from one individual to another. Even when they can perform written calculations, they often have trouble using those math skills in daily life situations.

RECEPTIVE LANGUAGE UNDERSTANDING

Students with autism spectrum disorders usually have great difficulty understanding spoken language. It seems that when they are young, they have a much longer processing time for oral language, so they may easily fall behind in comprehending what has been spoken. Following directions, even with the use of manipulatives, can be difficult and slow.

Children with autism do not have the advantage of most children with Down syndrome—that of being able to learn by reading the speaker's body language. Children with autism have a great deal of difficulty understanding the gestures and the implicit meaning behind spoken language. They tend to see the meanings of words very literally. For example, the teacher may say, "*Match* these two pictures." The child with ASD may have learned that the word *match* means something that makes fire, especially if she has been learning vocabulary from flashcards. She may be confused and look for a fire-producing match. Or a student with ASD might associate the word *time* with clocks and not with a way of describing the process of multiplying.

IMITATION OF OTHERS

Young children on the autism spectrum, including those with Asperger syndrome, do not automatically imitate other children. Often they need to be taught how to imitate others. Many older students with autism have learned the coping device of looking at another student to get the cues for what the teacher expects. Therefore, these students can benefit from working openly with manipulatives in a general education class.

NEED FOR VISUALS

Students on the autism spectrum can often be taught effectively using concrete objects or visual representations of things that they do not understand in oral speech. Visual schedules and tasks that can be broken down to small, sequential steps and pictured by illustrations or pictorial symbols such as the Picture Communication Symbols© created by Boardmaker software are used by most successful teachers of students with ASD.

Students with autism may also need assistance progressing from representing math concepts with concrete objects to using the abstract symbol system that we use to refer to mathematics. Even if the numerals' names come easily to them, they may still have trouble with abstract vocabulary terms such as *subtrahend, tens* vs. *tenths,* and *associative* and *distributive* properties. These students may need a great variety of experiences and materials to make the transition to meaningful abstract math performances.

NEED FOR EXPLICIT TEACHING

Students along the autism spectrum often need to be explicitly taught some social and other skills that other students just absorb naturally. They may have trouble understanding story or word problems because they may be based on unfamiliar social situations. For example, a student might be expected to solve story problems that are centered on preparing and giving a birthday party for a friend. If the student does not have experience with having a friend or being part of a birthday party group, she may not understand the problems. Or the problems may focus on giving Christmas gifts, but the student doesn't understand the hidden assumption that when someone gives a gift to you, they may also expect a gift back. Teachers and parents should be aware that social skills difficulties could underlie a math difficulty and be prepared to explain the social situation as part of the story problem.

Learning Characteristics of Students with Learning Disabilities

Students with learning disabilities have more difficulty learning skills in a particular academic area than would be predicted based on their overall cognitive abilities. Usually, students with learning disabilities (LD) are diagnosed with reading problems. However, authorities estimate that 6 percent of all students have significant math deficits—with a much higher percentage of students with reading disabilities also having math difficulties (Garnett, 1998). Learning disabilities in math are technically known as dyscalculia. Math difficulties range from mild to severe and vary in type. For instance, some students with a math learning disability just have trouble memorizing math facts but can understand higher level math if allowed to use a calculator. Other students struggle with basic number sense, such as understanding how to count or which number is greater than another.

MEMORY DIFFICULTIES

One common difficulty is in memorization of the basic facts in all operations. Some students count on their fingers or use tally marks way beyond their early elementary years. Memory aids, often concrete or pictorial, can be helpful for these students. Some may never master the basic facts, however, and will benefit from using a calculator.

VISUAL-SPATIAL AND ORGANIZATION PROBLEMS

Other students with learning disabilities have visual-spatial difficulties along with troubles with organization that make pencil and paper math confusing. They may have trouble writing recognizable numerals. They may also have difficulties lining up numbers with similar place values, so they may, for example, struggle to get all the tens lined up in a column. Manipulating objects can be the key to helping these students make sense of mathematical concepts.

ABILITY TO FOCUS ON LEARNING

Some students with learning disabilities also have attention-deficit disorder (ADD) or attention-deficit/hyperactivity disorder (AD/HD). Students with AD/HD or ADD often need their learning sessions to be of shorter duration. However, it is quite possible that students with AD/HD could play games and focus on concrete activities from start to finish. Russell Barkley, a leading authority in the area of attention-deficit disorders, has noted that many children with AD/HD or ADD can play video games for long periods of time, probably because the games give immediate feedback that keeps them involved. Barkley concludes that the most important factor in AD/HD is not a deficit in attention but the impulsivity that the child shows and the fact that she is easily bored. Interesting games and concrete activities could therefore be a welcome addition to traditional math teaching for students with ADHD.

Students Who Have Had Insufficient Math Experiences

Many students, with or without the disabilities discussed above, have not had the instruction and/or experiences needed to help them make sense of abstract concepts in math. This may be because they are young and simply have not had enough practice relating informal math experiences at home with the abstract math symbols used to represent them. Then again, this may be because they have not been given sufficient opportunities to experience the real-world uses of math—perhaps because their families have not made this a priority or because adults around them have not had high enough expectations about their math learning abilities. Some of these students may have been formally taught about abstract concepts and skills, but because they were not engaged by the teaching methods, they tuned out and decided math wasn't for them.

All of these students need frequent experiences with objects that can be moved and manipulated, as well as many opportunities to hear the labels that are used to describe them mathematically. Then they need help making a careful transition from the concrete objects to pictorial representations and then to the abstract symbols and vocabulary that are used for math. In fact, most concrete learners with any diagnosis need careful teaching from concrete to semi-concrete to abstract math symbols.

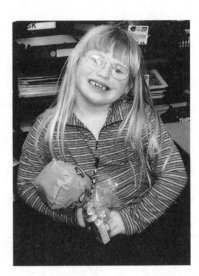

Conclusion

Whatever the reason for a student's difficulties in math, it never hurts to try hands-on activities with the student, whether or not you know for sure that she is a concrete learner. This is especially true if the student is under 11, because most children that age are still concrete learners. However, the same may be true of students with Down syndrome, autism spectrum disorders, or cognitive disabilities, as well as older students who have learned little in traditional math instruction. The most important key to meaningful math learning seems to be the progression from Concrete to Semi-Concrete (Representational) to Abstract, as shown below.

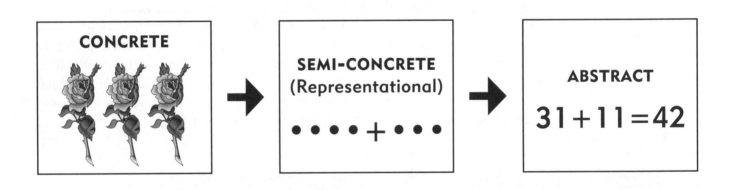

CHAPTER 3

Teaching Strategies

Questions to be answered:

1. What teaching strategies and procedures are successful with concrete learners, including those with Down syndrome?
   - Emphasize visual learning
   - Use hands-on activities
   - Provide structured learning with some flexibility
   - Make learning relevant to the real world
   - Focus the student's attention
   - Provide nondistracting written work
   - Give simple, clear homework
   - Minimize fine motor demands
   - Expect and encourage appropriate behavior
   - Ensure early success in the lesson
   - Consider the use of peer tutors
   - Facilitate short- and long-term memory
   - Use the calculator early and frequently
   - Make your interactions enjoyable
   - Use the computer
   - Break down the task into small steps

2. What adaptations and modifications can be made to general classroom work?

Many of the teaching strategies that are very helpful with children who are concrete thinkers are simply good teaching techniques that work with most students. If a teacher of general education classes is systematic, creative, and tuned-in to the learning styles of all the students in her class, she will probably have the skills that she needs to work with students who are concrete thinkers. Of course, it helps to be a "ham" and a good storyteller and to be enthusiastic about learning, too.

Even though I have been working with students who are concrete learners for years, I still occasionally look at some lists that I have compiled to help me remember good strategies for teaching that just may work with that special person. I am including the strategies and procedures here that I included in *Teaching Math, Book 1* that seem to work best with students who are concrete thinkers. If you have already learned this information from Book 1, you may skip this chapter. I will also discuss some adaptations and modifications that can be made for these students in the general education classrooms.

Teaching Strategies and Procedures

EMPHASIZE VISUAL LEARNING

Students with Down syndrome, students with autism, and other concrete learners are very often visual learners—that is, they learn better by seeing actual objects or pictures of concepts rather than hearing someone talking about those concepts. Time after time, I have seen students' eyes light up when I show a picture of what I am explaining. Most elementary teachers often use visual aids (commonly called visuals) in their teaching. However, when a student goes to middle school or high school, people seem to assume that reading a text should be enough. Visual learning through pictures and graphs are important to all of us as we are learning, regardless of age. In addition, because today's students spend so much time looking at television, playing video games, and using computers, many of them have come to expect visual explanations for learning.

To make learning visual, it is not sufficient to just write numbers on the blackboard or on an overhead projector as you are talking. For younger children, using objects or pictorial representations of activities will be the most effective way of teaching. Once a child is able to read fluently, written numbers or words can supply a lasting explanation that is useful. (Visuals can be looked at repeatedly to help with working memory.) However, even older students may be able to learn more quickly when pictures are included with words or numbers.

Educators working with students with autism spectrum disorders (ASD) have recently developed good innovations in the use of visuals for academic learning, direction following, and social skills. For example, Carol Gray has come up with the concept of "Social Stories"—illustrated stories written by parents, teachers, or language therapists tailored to specific social problems a student is having that can be read and reread to help his understanding. And many students with ASD learn new skills through activity schedules, which are sequences of photographs or drawings that illustrate the steps involved in the skill. Many of the materials designed for students with ASD can be used as-is or modified for use with other students who are concrete learners. Some useful books related to visual strategies are listed in the Resources at the back of the book.

USE HANDS-ON ACTIVITIES

By definition, students who are concrete learners learn by manipulating objects and working out solutions with hands-on activities. Sometimes materials are left out for the students to discover their properties, but more commonly the parent or teacher demonstrates the actions and the student reproduces the activity with his own materials. These hands-on activities may be more difficult for the parent or

teacher to orchestrate, but the student's learning is usually greater and longer lasting.

Think about how you learn. Are you one of those people who just can't make heads or tails out of the instructions in your computer program manual? But let someone show you and guide you in doing the steps yourself, and those actions make sense, and you can do them easily. For me, demonstrations are often not enough. I watch the instructor doing origami and it looks easy. However, when I try to make those folds myself, I need direct help in the folding.

Students who are concrete learners need demonstrations and hands-on activities for almost everything they learn. Math is often a rather abstract process as taught above fourth grade level. Concrete learners need activities that make the process real for them much more than they need pages of written problems. Every time you introduce something new, you should be thinking about how you could help the student to do an activity that will help this new idea be understood. Although many activities you may come up with involve fine motor skills such as manipulating beans or counters, it is also important to devise some activities that can be done with large muscles. For example, students can move to specific areas of the floor to do sorting activities: "All students whose favorite color is red, move to the front door. Now let us count out loud how many people love red."

Another large muscle activity could be used to teach and practice the recognition of numerals: "Put your foot on the numeral 2" shows whether the student recognizes the numeral 2. Even the game of hopscotch can be used for number recognition.

PROVIDE STRUCTURED LEARNING WITH SOME FLEXIBILITY

Many students who are concrete learners are more comfortable with having a schedule and following it faithfully. They would like to have things stay just as they usually do it. As teachers, we need to address their need for structure and also build some flexibility into their learning.

For younger children, we can warn them that in five minutes we will be switching activities. Sometimes this warning is not enough, and we will have to think up a leadership-type activity to help them transition into the next subject. "Tom, I need you to collect the scissors for your group."

Making a classroom schedule with every activity on a separate card (illustrated, of course) and posting it in the front of the room may be useful. I put magnets on the back of large cards so they can be used on the refrigerator or a metal-backed chalkboard. Then if a shift in sequence is necessary, you can make the change as the child watches. For some students, it may be helpful to have their own individual schedules for each day of the week. These schedules can be put in a small pocket chart holder so they can be changed. Parents working at home on math should also have a visual schedule or routine that helps their child know what to expect. Routines are especially important for students on the autism spectrum, who often have much difficulty with change.

Although schedules are useful, it is also important for children to learn to deal with unexpected changes to the schedule or routine because in real life, things don't always go as planned. One teacher in a workshop told me that she had one card with a question mark on it that she used when plans had to change suddenly. Sometimes the card signaled a nice surprise or a treat so the students had a pleasant experience with change.

MAKE LEARNING RELEVANT TO THE REAL WORLD

Just because you have the visual symbol for math at a particular time of the day, it doesn't mean that that is the only time you work on math with the student. It is very important that you weave the skills you are teaching in math into the other parts of the academic day and into real-life settings. Involving students in activities such as measuring sugar for a cake or buying a video with their own money helps them understand why they need to learn something and motivates them to learn.

Parents often have the most opportunity to make math instruction come alive. Grocery shopping, mall shopping, counting calories, using the phone, punching in

numbers on the TV remote, going to a restaurant, and many, many other activities at home and in the community can provide opportunities for math learning and practice and motivate your child to use math.

Be careful, however, that the real-world activities are not too complex for your student. Ads and signs often include numbers that can be confusing to students who are learning about money. For example, prices that are given per pound can give the student the idea that the entire piece of meat is only $.99. Or a grocery newspaper ad may offer 3 oranges for $1.00 when the student has not yet learned the process of division.

FOCUS THE STUDENT'S ATTENTION

Some parents and teachers have said that "I could teach him if I could only get him to sit still long enough." Getting the student's attention may be difficult for several different reasons. First, if he is young, he is at the stage where he should be exploring and going from activity to activity. How long does a typical child sit to do activities at the same age? Are you expecting too much?

Another factor may be that you are not matching your teaching to your child's interest level. If the activity is dull or your voice is whiny or monotone, you may have lost him at the first turn of the race. Are your words at his level? Are your sentences fairly short? Are you stating the information clearly? Are you enthusiastic? Is the information concrete enough? Do you have pictures or models that are hands-on?

In addition, you may need attention-getters just to awaken a student's interest. For instance, when beginning to teach about measurement, you can do something unexpected such as coming into the room with a ruler balanced on your head but acting as if you don't know it is there.

Most typically developing older students can motivate themselves to do assigned work. Older students who are concrete, hands-on learners may not make the connection that doing work now will help them get good grades later on. You may have to set up some type of reward system such as putting a sticker on their paper if they started to work in three minutes. The reward must be immediate or part of token system until they have enough experience to want to do well just for themselves (internal rewards).

Finally, some students with Down syndrome, autism spectrum disorders, and learning disabilities actually satisfy the diagnostic criteria for attention-deficit/hyperactivity disorder (AD/HD). If your student has a dual diagnosis, you may need to read books or attend workshops designed to teach strategies for focusing and keeping children with AD/HD on task. Essentially, the principles behind keeping attention on task are to give the student almost immediate feedback and to make sure the tasks are varied and interesting.

PROVIDE NON-DISTRACTING WRITTEN WORK

The written material that is given to the student should not distract from the purpose of the activity. Sometimes artists of school materials want to make them more aesthetically pleasing, so they use a special color or design to highlight items. These additions may be distractions for the concrete learner.

Sometimes there are just too many pictures in one area. I found that the Richard Scarry books that I had read and looked at with my other children were just too busy for my son with Down syndrome. Watch the student's eyes. If they dart back and forth or if he is unable to locate a prominent feature on the paper, take a look at the materials and see if they have too many distracting elements. One thing you can do to minimize distractions is cut a hole in a sheet of paper, cover the picture, and show him only one part of it at one time.

If you are preparing written work for the student, use your computer or copy machine to make things larger, even if it takes two sheets instead of one. It is better

to have fewer math problems on a page and leave more space for the student to write in the answers. Make sure that the print quality is dark, crisp, and legible, so that the student does not have to guess whether a number is a seven or a one, or whether he is looking at a picture of a nickel or a quarter. Some authorities suggest that text writing should be done in 14 point and Arial font.

A study done in the United Kingdom (Wolpert, 1996) with ninety teachers who had students with Down syndrome in general classrooms found that workbooks were not useful at all. The authors felt that the workbooks were either heavily dependent on language comprehension or had too many distractions or problems on a page, which made it confusing for students with Down syndrome. The teachers felt that the most effective materials for instruction were concrete objects or manipulatives that the student had to use to perform an activity.

Don't give too many directions at the same time when explaining how to do the written work. Processing multi-step directions can defeat a student at the very start. Write the instructions on the worksheet, especially if it is going home. Parents will often help at home, but not if they can't understand what has to be done.

MINIMIZE FINE MOTOR DEMANDS

Fine motor problems can make it harder and more tiring for children with Asperger syndrome and Down syndrome to hold a pencil and form numbers. When their hands get tired, they can rapidly lose interest in a written assignment. For this reason, you should minimize the amount of copying (from the board or book) you ask the student to do. Whenever possible, it is preferable for an adult to copy down the problems or to use photocopied pages that the student can write on.

When it is essential for the student to copy problems himself, be aware that students with Down syndrome and students with learning disabilities often have problems with writing down number problems. The ones may be put under the tens column, the numbers may not be recognizable, or the decimal points may be lost. You can use large graph paper so that the student writes one digit in each square and lines them up. A simpler solution can be to turn lined paper sideways so that the lines are vertical (running up and down). The student can then put the numbers in columns. A computer program, *Access to Math* (Mac only), from Don Johnston makes worksheets for you where the numbers are always lined up and shaded areas tell you where the totals should go. (*Access to Math* also prints a worksheet with the right answers for the teacher or parent.) Some math worksheets can be created and printed from the website SuperKids.com (www.superkids.com/aweb/tools/math).

GIVE SIMPLE, CLEAR HOMEWORK

Homework sometimes becomes a source of friction between school and home. The work sent home for math homework should always be something that the student has already learned. Parents should be reinforcing the school teachings, not teaching the student what he didn't pick up in school that day. If the concept has really been learned in class, the student should be responsible for doing some of each type of problem, but not necessarily the same amount of work given to typical students. Students with fine motor difficulties who have to spend excessive amounts of time writing recognizable numbers should be allowed to do only a representative number of problems, less than the rest of the class.

If a student has really worked hard at a subject for twenty or thirty minutes (depending on his age) and doesn't seem to understand the assignment, Mom or Dad should have an agreement with the teacher that he can stop work. They should then send a note explaining that the student has worked hard for twenty or thirty minutes straight and still does not understand. Ask the teacher to explain and demonstrate again to the student what is expected. This will lower the frustration level at home—for student and parents. Parents have to resist the temptation to do the problems for their child. The teacher will not know what areas need to be re-taught if he brings in work he hasn't done himself. (However, it may be acceptable for you to write down your child's answers for him if his hand gets too fatigued.)

Parents can help by having a consistent time for homework that doesn't compete with favorite TV shows or outdoor play. Homework should not be a constant time for conflict between student and parents. If homework is creating such trouble at home, parents, teacher(s), and the student need to talk together and work out some solutions.

Rushing out the door, Mrs. Landon said, "I have to go home now. It is Heidi's homework time, and nothing happens if I am not there. I am just getting to hate the time after supper at my house. I don't know how much help to give. Sometimes she says, 'I can do it myself.' and sometimes she says, 'You never help me.'"

EXPECT AND WORK TOWARD APPROPRIATE BEHAVIOR

Sometimes it is the student's behavior that keeps him from learning. Appropriate behavior is not an area where we can give a few tips that will fit all students who are concrete thinkers. You need to step back, look at the disturbing behavior, and find out what is causing it. That may be something that happened immediately before he screamed and kicked, but you must also look at what has happened further back in time. Also important is to find out what function the screaming and kicking behavior serves for the child. (There is a whole philosophy of looking at behavior that is called positive behavior support, and one of its chief tools consists of doing a functional behavior assessment.) Then you have to look to see what happens to Charlie after his meltdown and see if his behavior is reinforced by the consequences.

For example, Charlie started screaming and kicking today when you closed his book and steered him over to the table for math. Closing the book and steering him over to the table was the immediate trigger for his misbehavior. However, we don't know why he did it. We don't know the function that the meltdown behavior serves for Charlie. Does he get total class attention? Is that what he is trying for? Has he been so frustrated with his last couple math sessions that he wants to get out of the whole situation? Does he get to skip math when he screams? Was he so absorbed in looking at the pictures in the book that he was really upset with having the book closed so abruptly? Did he have enough warning about the coming change? Does he have a schedule to follow? Has everything up to this time been off-schedule and confusing?

You can see that without knowing why Charlie is misbehaving and what he gets for the behavior, you can't work on his behavior in an effective manner. You need to look at:

1. What happens immediately before the misbehavior?
2. What happens to Charlie when he misbehaves in this way?
3. What has Charlie's experience been with the math activity for the last week or so?
4. What function does the misbehavior serve for Charlie?
   - Does he get attention from the class, parent, or teacher that he enjoys?
   - Does he get out of doing the task he has been given?
   - Does he get to return to a more pleasurable activity?
   - Does his behavior stop a physical sensation he does not like (such as the teacher touching his shoulder)?
   - Especially if he has an autism spectrum disorder, does moving from the floor to the table stop some relaxation movements that he needs to be able to focus?
   - How can we satisfy the need that Charlie is showing with his misbehavior in a positive way?

A study was done in the United Kingdom (Wolpert, 1996) which surveyed 90 parent/teacher pairs about the inclusion of students with Down syndrome in regular classrooms. The students had an average age of 9.3 years and a grade level of 3.8. When asked about behavior management strategies used in the classroom, participants said the most successful method was praise from the teacher. Threat of lower grades was not an effective motivator for students with Down syndrome to work harder. Punishment and ignoring inappropriate behavior were also considered ineffective for behavior or instruction. The researchers felt that the children were probably not able to connect their behaviors with the consequences.

The teachers felt that the most effective way to correct behavior was to calmly point out what behavior was not appropriate and the consequences. Consistent follow-through on classroom rules was also essential.

Token systems based on positive actions have been successful for many students with learning difficulties, but may work best if the whole class is on the system. Token systems may also work in a family, where the children get tokens such as stickers for desired behaviors and receive some reward at designated times. Privileges can often be used as rewards more easily in families than in classrooms because of school rules.

When working one-on-one with a student, I sometimes got a frown, a head down on the table, and a refusal to do any more. I might set a puppet up on a pop bottle and give the instructions to the puppet and laugh and tease it just as if the puppet was another student. Often that is enough to get the student to look up and try. Sometimes I have to say something that is absolutely wrong, and the student can't resist correcting me. My son and his buddies (all young adults) can be kidded about being grouchy. I imitate what they look like (exaggerating, of course), and they usually start giggling and get back to work.

Working with a student at home does not always bring out the best behavior in him. Sometimes children have a difficult time interacting with mother or father as a parent and as a teacher. This is especially true if one parent does all the teaching. It does get frustrating when your child can't remember something that you thought he understood perfectly yesterday. It is common for irritation to creep into your voice, although you may not be aware of it. If you find yourself feeling irritated frequently, your child will probably respond better if you bow out and let the other parent do some of the teaching.

Other children in the family can also help for short, directed lessons. You need to monitor their teaching so it is positive and directed toward the proper goals. I recently came upon some notebooks I had written when my son Scott was younger. I had detailed directions for my daughter Heidi to give lessons on sorting and matching to Scott. I remember that those directions were written one summer that I gave my daughter the equivalent of the pay she would have made in a fast food job in return for babysitting and teaching Scott. She learned teaching skills and got closer to her brother Scott. Scott used to say that he had a mommy and a daddy and a Heidi—which he thought of all in a similar way.

It is possible that your child just will not accept you as a teacher. That is all right. Your most important role must be as a parent. You may be able to find a tutor or an interested neighbor who can teach your child using this book. You also might be able to arrange some individual time with a teacher to use the one-on-one activities given in the book. I usually put on a funny-looking hat when I worked with Scott as a teacher. I told him to call me Mrs. Horstmeier, not Mom, when I was being his teacher. For academics, this seemed to work well. However, when I tried the same approach to teach him to play the piano, it really bombed! Luckily, I found a piano teacher who stepped in successfully.

ENSURE EARLY SUCCESS IN THE LESSON

Success is very important at every level of teaching for students with Down syndrome. The steps in teaching should be very small, and each one should be praised for effort, if not correctness. As the students get older, they often meet with frustration as they try to do what others do. They may give up very soon because it is better to not do the activity than to do the activity and fail. Sometimes the students who are "higher functioning" are more aware of those discrepancies than those who have more significant cognitive disabilities, and, consequently, feel more stress about the situation.

I have tried to include a "success step" before each major new learning in this book. That is, I recommend a relatively easy activity related to the new skill for the student to complete at the beginning of the lesson. It will only be a success step if the student is able to do it easily. Do not hesitate to change the step to something you know that the student can do well. You may find that some of the activities jump too quickly from one step to another. By observing the student, you can tell whether an intermediate step is needed and can act to preserve the feel of success. Don't hesitate to try your own intermediate step.

Of course, we do not want students so dependent on praise that they can't function without it. In the 1970s, many professionals went overboard on giving praise. "Good talking, good creeping, good eating…" could be heard from many preschool and elementary school classrooms. I felt my son got to be a "junkie" on praise for everything. Some of this praise can actually present a barrier for meaningful communication. If you are saying, "Good talking," you are not responding to the content of the student's message in a meaningful way.

As we get older, we are able to see our own accomplishments and get satisfaction from them. We learn that in most work environments, "No news is good news." But even adults need occasional quiet praise for their self-esteem. So, we should sincerely praise our students but also teach them to feel good about their own success. When Scott was younger, I would say to him, "Don't you feel good when you can do that correctly?" Often now, he says quietly to himself, "I did it!" when he has accomplished something.

CONSIDER USING PEER TUTORS

Both parents and teachers in the UK study mentioned above felt positive about peer tutors. In fact, the teachers of sixth grade and higher reported that peer intervention worked better than teacher assistance. In the upper grades, students are striving for more independence, and being helped by a friend was perceived as better than being helped by a teacher or aide.

However, peer tutoring should be organized with care. When the students are in first grade or below, they should be just helping each other, not having a peer-tutor relationship. When a peer is actually doing tutoring in an academic subject, the teacher should carefully select the tutor and actually teach him or her some teaching techniques. See if you can enlist a "laid-back" student with a sense of humor and the ability to follow directions. I have the tutor observe a session when I am working directly with the child to model the teaching strategies.

After the first one or two tutoring sessions, the teacher should talk to both students individually about the relationship and the skills learned. The tutor may be puzzled about problems he or she has encountered, and the tutee needs to know that he is expected to study and work with his peer tutor. It is important to look at what goes on in the tutoring session. I have seen sessions where the tutor makes fun of the student he is supposed to be helping, which is exactly the opposite of the effect we are hoping to create.

If you are working at home, it might be helpful to enlist a sibling as a classmate and even have the sib do a little tutoring. My two oldest children did a good job of tutoring Scott, but his brother just three years older was too caught up in sibling rivalry to create a positive atmosphere. He did, however, enjoy playing math games with Scott.

FACILITATE SHORT- AND LONG-TERM MEMORY

Assist the student to facilitate short- and long-term memory storage by making musical or rhythmic associations with the concepts, using concrete visualizations, encouraging creative practice, and using mnemonics.

Music is an excellent way to help students memorize numbers and important concepts. Many children (and adults) can remember the tune and the lyrics to a song that teaches a concept. Parents and teachers can make up teaching verses to many common childhood songs. It is important to remember that the tune should be simple and easy to sing so that the emphasis is on learning through the words. A chanting rhythm can also be used to help memory storage. For example, Jan Semple, in her book, *Semple Math* (1986), has the students chant together while tapping pointing fingers, "Numbers together go plus, plus, plus," as they make the plus sign with their fingers.

Concrete visualizations of objects and pictures help students make a visual picture in their minds. We found that many students needed to see the passage of time on a round analog clock even though they could read the time on a digital clock. Likewise, many people picture a cut-up pie or pizza when they think about fractions. Actual manipulation of the materials in problem solving adds the tactile sensation of touching to the visual learning.

Mnemonics, making picture associations with the numbers and concepts, is also a helpful strategy. For example, we easily remember the shape of Italy because we have learned to associate its looks with a boot. Some older school classes share mnemonic strategies that help them remember important items. When reading a word problem that included the word together, one student observed, "Together starts with t and that looks like a + (plus) sign for addition."

Short reviews at the beginning of each class period will help to make important facts automatic. Practice also should be introduced in a fun way with varied and creative activities. Ideally, each student should learn the addition and subtraction facts from 1-20 by heart, although this is not an essential skill for survival. That way, simple calculations can be done without paper, pencil, or calculator. Most individuals with Down syndrome will need to practice and practice the simple addition and subtraction facts if they are to be useful in this way.

USE THE CALCULATOR EARLY AND FREQUENTLY

When pocket calculators first became widely available, typically developing students weren't usually allowed to use them much before middle school, on the grounds that early calculator use might interfere with learning math facts. However, an analysis of many research studies (Hembree & Dessart, 1986) has concluded that use of a calculator, along with traditional math instruction, improved the average student's ability to do pencil-and-paper calculations and to problem solve. The National Coun-

cil of Teachers of Mathematics (1989, 2001) has suggested putting less emphasis on paper-and-pencil calculations and using more technology.

I have noticed that many adults who are concrete learners do not use much mental math in everyday situations, perhaps because of working memory problems. They can, however, be taught to use calculators well. If some individuals with cognitive disabilities, especially those with Down syndrome, are not able to use mental math in a practical way, why not teach them to use the calculator from a young age? You can have the students use the calculator to check what they are doing with the concrete materials, and checking the calculator answers with concrete materials. Use of the calculator should be automatic for any appropriate math situation.

MAKE YOUR INTERACTIONS ENJOYABLE

Make your teaching fun and indicate your pleasure in working with your students. Often they can be motivated by mild competition. If there are no other children present, challenge the student to compete with you. Of course, you will not always win, and you are so surprised that he has learned so much. Use humor as much as you can.

USE THE COMPUTER

Most computer programs are not designed to teach math but instead provide practice for concepts that have already been learned. The color and graphics in these programs usually get the student's attention immediately. Then the immediate feedback for right or wrong answers can keep him on the learning track. The student can usually learn to use practice-type programs in one teaching session, and he can work independently from then on. Working with devices such as the calculator and the computer seems to be intrinsically interesting to students who are concrete learners. They see color and graphics in front of them and they get immediate feedback on their answers.

Other computer programs do some teaching, especially if the student is told the correct answer after one or two tries. The computer is definitely not a substitute for a teacher, and it takes a teacher to set up the situation so that the computer session is productive. However, allowing the computer to do the teaching now and then can be a welcome change for both the teacher and the student, as well as being good motivation for needed practice. Computer software is continually changing, but a few good programs are listed in the Resources section.

BREAK THE TASK DOWN INTO SMALL STEPS

I have tried to break down the goals and tasks in this book into small steps that can be mastered easily. However, every student will learn in his own way. I'm sure that you will find times when this particular student just can't make the next step in the math goals. You will have to find a way to make the task shorter or less complex or try a completely new activity. Most parents and teachers who know how a given student learns best can find a simpler or more effective way to teach the concept. Try it. Breaking down or modifying a task is a lot easier than most people think.

Adaptations and Modifications to General Classroom Work

In the early elementary years, it is possible that some students with Down syndrome and other concrete learners may be able to do exactly the same work in math class as the other students. By third or fourth grade, however, almost all students who are concrete thinkers will need adaptations in the way math is taught or in how the student shows his learning. There will probably need to be some modifications in the content of what the student will be learning as he gets to the higher grade levels.

It is beyond the scope of this book to go into great detail about ways to make math work more appropriate for students with disabilities. I do want to stress the importance of determining the most appropriate adaptations for your child and spelling them out in his individualized education program (IEP). With increasing numbers of children with learning problems being included in general education, you cannot count on your child's teacher having the special education training to know how to modify classroom demands for your child. There are books dedicated to helping teachers modify the curriculum in the general classroom (Blenk, 1995; Hammeken, 1995, Beninghof, 1998; and Stainback & Stainback, 1992). (See the References.) One of the most frequently mentioned programs is explained in the book, *Adapting Curriculum and Instruction in Inclusive Classrooms: A Teacher's Desk Reference* (Ebeling, D.G., Deschenes, C. & Sprague, J., 1994) from the Institute for the Study of Developmental Disabilities, Indiana University. This combination book and staff development kit categorizes nine types of adaptations:

1. Size—reduce the number of items
2. Time—extend amount of time for test or assignments
3. Level of support—provide more assistance
4. Input—modify the way the instruction is given to the student (for example, read the problems aloud to him, or provide manipulatives)
5. Difficulty—make the problems easier (for example, by using a calculator or simplifying the rules of a math game)
6. Output—adapt how the student reports his learning (for example, using stamps or labels with numbers printed on them, rather than writing them, or having an aide write down the student's answers)
7. Participation—the student participates in only part of the task (for example, the student could gather data about favorite ice cream flavors with the other students, but then not figure out what percentage like vanilla best)
8. Alternate goals (modifications of classroom goals)—have less complex goals than the rest of the class (for instance, learning single-digit subtraction instead of three-digit subtraction)
9. Substitute curriculum and goals—student has different instruction and activities for his specific goals

Because I have found that I often need reminders of each student's special needs, adaptations, and modifications, I have compiled a checklist that can be filled out at the beginning of the year for each student and given to all other teachers involved. I have modified the checklist to be more specific to mathematics education.

Parents can send the checklist to their child's teacher or bring it to the IEP meeting and discuss the most appropriate adaptations to help their child. You have the most accurate, complete knowledge of your child and can be of real assistance to the school.

Student Name: _____ Date: _____

Checklist for Adaptations and Modifications to the General Curriculum

The following adaptations are appropriate and necessary for this student. Check all that apply.

Pacing
___Extend time requirements
___Vary activity often
___Allow more breaks for student
___Omit timed assignments
___Work on vocabulary before lesson
___Pick out only major concepts for learning

Environment
___Reduce/minimize distractions
___Provide extra paper and pencils close to student

Presentation of subject matter
___Teach to student's learning style
  ___Visual
  ___Auditory
  ___Tactile-kinesthetic
  ___Experiential
    Such as:
- Use visual whenever possible
- Use visually colorful computer programs
- Use pictures and mnemonics for memory
- Use chants or songs
- Use sand or Kool-Aid in a pan for writing
- Use manipulatives and hands-on activities
- Write with finger on desk when learning
- Do wet writing on chalkboard
- Use dry erase marker on whiteboard
- Practice with board games
- Reduce copying from the board/book

Type of instruction
___Individual and small group instruction
___Functional application of academic skills
___More review
___Move around the room to gather information
___Errorless learning

Materials
___Large print
___Arrangement of nondistracting material on page
___Calculator for all math
___Graph paper
___Computer (not just as reward)

Assignments
___Visual daily schedule
___Calendars and assignment books
___Use written back-up for oral directions
___Request parent reinforcement
___Reduce paper and pencil tasks
___Shorten assignment
___Lower difficulty level

Testing and proof of learning
___Provide thorough reviews before tests
___Oral testing
___Correct missed problems for extra credit
___Test administered by aide or special ed person

Social interaction support
___Peer advocacy
___Shared experiences in school
___Extracurricular activities
___Structure activities to foster social interaction
___Train peer tutors
___Debrief peer tutors

Motivation and positive climate
___Offer choice
___Planned motivating sequence of activities
___Mostly positive reinforcement
___Verbal praise
___Concrete reinforcement, if needed
___Set up token system
___Use strengths/interests often
___Cultivate a general positive attitude

Individual hints for working with this student:

CHAPTER 4

Informal Assessment

Questions to be answered:

1. How do you decide where to start a student in *TEACHING MATH, BOOK 2*?

2. How do you give an informal assessment?

3. What is assessed by the game Earn & Pay?

4. What is assessed by the game The Journey?

5. How do you assess functional skills such as banking and shopping?

6. How do you score the two game assessments if you need to have a record of achievement?

How Do You Decide Where to Start a Student in This Book?

You can determine which math activities to use with a student you are just beginning to work with in several ways:

1. If you have been doing math with the student using *Teaching Math, Book 1,* or some other program in a small group or one-to-one situation, you probably know just what instruction that student needs.
2. Use the Pre- /Post-Checklists at the beginning of each instructional chapter.
3. Do the Informal Assessment in this chapter.

Informal Assessment

Computation skills can be assessed by having the student play two games. The informal assessment is very useful if you are not sure what the student's skills are. It is called an *informal assessment* because it is given for diagnostic purposes only. Therefore, no age norms or grade levels will be given.

EARN AND PLAY GAME

Teaching Math, Book 1, introduced the game Earn & Pay as a means of assessing math skills. In the course of the game, students earn money and pay for items as they progress around a game board. Earn & Pay can also be used as an assessment for *Teaching Math, Book 2.* The following are areas that can be surveyed by having your student(s) play the game:

Simple money concepts
1. Does the student understand the concepts of paying and receiving money?
2. Can the student skip count by 5's and 10's?
3. With the piles of sorted currency in front of them, can students skip count the $5 and $10 bills in each pile correctly?

Simple multiplication
4. With the piles of sorted currency in front of them, can students multiply the value of the bills times the number of bills?

Next highest dollar strategy
5. When the student has to pay for something, does she give the next highest dollar or dollars and expect change back?

Simple subtraction
6. Can the student tell you how much money she should get back when she gives you the next highest dollar? She may use the calculator.

Simple addition
7. Can the student add two or more amounts to get the total amount of money she has left? She may use the calculator.

ADVANCED SURVIVAL SKILLS ASSESSMENT— THE JOURNEY GAME

Teaching Math, Book 2 includes instruction in more advanced computational skills than Book 1 does. The Journey Game assesses those skills by a game in which players travel to the top of a mountain while solving math questions from cards handed to them by the assessor. The instructor chooses the types of computational and words problems that are given to each player. The skills assessed are:
- Addition number problems
  - Single digit
  - Multiple digits
- Addition word problems
- Subtraction number problems
- Subtraction word problems
- Multiplication number problems
  - Single digit
  - Multiple digits

- Multiplication word problems
- Division number problems
- Division word problems

Instructions for Games

INSTRUCTIONS FOR EARN & PAY

MATERIALS:
- The Earn & Pay game from Appendix B.
- Earn & Pay cards copied on cardstock from Appendix B.
- Money (play) $1, $5, $10, $20, and $50 bills, enough for $85 for each person. (Smaller versions of currency, photocopied only on one side, are available in Appendix B, although realistic play money is preferred.)
- Money Total slip from Appendix B
- Dice
- Game markers (small buttons, tokens, pennies, etc.)

PROCEDURE:
1. Copy the game board from Appendix A and paste the pages on the inside of a file folder.
2. Cut out the Pay and Earn cards. Photocopy the Pay cards on one color of paper and the Earn cards on another color, or paste them onto different colored construction paper. Put the cards face down on the game board as indicated.
3. Give each player $85.
4. The first person rolls the die and moves her marker that number of spaces. The space landed on will be either a Pay space or an Earn space. The player picks up the top card from either the Pay pile or the Earn pile and receives or pays the amount listed on the card.

Sample of Money Total slip

Name: _____

Name of bill	Number of bills	Total amount of money in this bill
$1.00	**13**	**$13.00**
$5.00	**3**	**$15.00**
$10.00	**2**	**$20.00**
$20.00	**2**	**$40.00**
($50)		
	Final Total	**$88.00**

5. The corner squares on the game board involve some kind of direction, such as lose a turn or go back 2 spaces. If a player moves up or back following one of those directions, she does *not* pay or earn what is on that square.

6. The first person to reach the finish line is Winner I. She must throw the correct number to land on the Finish Square.

7. Winner II is the person who has the most money left. The students total up their money by using the Money Total slip. That requires them to count the number of bills in each denomination, multiply by the value of those bills, and then add those totals to find the final answer. They may also skip count the money they have left to figure out their total.

INSTRUCTIONS FOR THE JOURNEY GAME

MATERIALS:

- Journey game board from Appendix B on a file folder or cardstock
- Problem cards for addition, subtraction, multiplication, and division—both as number problems and as story problems
- Answer sheet
- Die
- Game markers for each player

PROCEDURE:

1. Place the game markers on Start. One student throws the die. She moves her marker the number of dots she has thrown with the die.

2. The facilitator hands the appropriate card to the student. If the student can solve the problem, with or without the calculator, she moves her marker one place ahead, if the problem is a number problem, and two places ahead, if it is a word problem.

3. The facilitator will give the student at least one number and one story problem for one-digit and two-digit addition, subtraction, multiplication, and division processes. If you can see that the student does not know multiplication or division, you do not have to continue giving those problems.

4. You can tell what operation is on the number problems by looking at the operations sign (addition +, etc.). You can tell which operation a word problem requires by looking at the middle letter in the code in the upper corner of the card—A for addition, S for subtraction, M for multiplication, and D for division. The first letter is just to prevent the student from figuring out the code so that she will figure out the proper operation for herself. The number is given so the student can find the correct answer on the answer sheet.

Administering the Informal Assessment

It is usually better to split the assessment into short periods over the course of two or three days. The actual board games should be played to completion so the students feel that they are playing real games. Much of the assessment information will come from observing the student. Each section is quite flexible so that the person doing the assessing does not have to use specific words or follow a specific order of items. Only the key areas in computation are assessed in this informal assessment.

Summary of Assessment:

Scoring for Pay and Earn	Correct	Points
1. Simple money concepts (paying and receiving)		
2. Skip counting by 5's and 10's		
3. Next highest dollar strategy		
4. Simple addition (figuring out total money left)		
5.Simple subtraction (figuring out change from bills)		
6. Simple multiplication (figuring value of bills times quantity)		
Note here any calculator use:		

Scoring for The Journey	Correct	Points

1. Addition in number problems
    Single digit
    Multiple digits		

2. Addition word problems		
3. Subtraction number problems		
4. Subtraction word problems		

5. Multiplication number problems
    Single digit
    Multiple digits		

6. Multiplication word problems		
7. Division number problems		
8. Division word problems		

Scoring

You really do not have to score this assessment, especially if you have only one student. If you note the areas where the student has difficulties, it may not be necessary to record any scores. If you work with several students, or if you need numerical scores to show growth, the scoring system is very simple. Just give similar questions to the student at the pre-test and post-test and compare the point value. If you are doing curriculum-based assessment, you will need to give probes during the instruction to monitor progress.

Point Values

Task(s)	Points
Student attempts task but does it incorrectly	1 point
Student does the skill inconsistently or prompted	2 points
Student does the task correctly after training* *Teacher demonstrates skill, gives the same item again (slightly modified by changing a number), and student succeeds. Usually the areas that the student can do with training are the first areas that you target for teaching because they will likely be the most successful.	2T points
Student does the task correctly without assistance	3 points

You can tell what computation skills the student needs to learn by comparing the assessment results with the names of the chapters and the pre- / post- checklists at the beginning of each chapter. If a student can't do any of the skills without assistance or after training, it would be best to go back to *Teaching Math, Book 1* and begin teaching those appropriate skills.

Assessing Functional Skills

Functional math skills related to handling money, shopping, budgeting, banking, baking with fractions, understanding graphs and charts, etc. will need to be assessed by the parents or other people who know the student well. You might begin with the following checklist. You can also compare the student's skills with the skills listed in the Pre- /Post-Checklists beginning in Chapter 5 or assess her skills through observation.

Checklist of Functional Skills

1. Can the student identify pictures of common fractions?	
2. Can she bake using common measuring spoons and cups?	
3. Can your student skip count 5's, 10's, and 20's as done in counting currency?	
4. Can she name the coins and their values?	
5. Can she buy an item using the next highest dollar strategy?	
6. Can your student accurately record the money she has received and spent?	
7. Can she fill out a check accurately?	
8. Can she budget a small amount of money?	
9. Can she explain how credit and debit cards work?	
10. Can she shop for and independently purchase more than one item?	
11. Can she figure out the price of several pounds of an item priced per pound?	
12. Can she compare prices on similar items?	
13. Can she figure an approximate state tax when given the total price?	
14. Can she identify right angles and parallel lines?	
15. Can she make a simple bar graph?	
16. Can she change a two-place decimal to a percent and reverse?	

Use the same scoring values as described above for the two game assessments. Again, if you need more detail on her ability to perform specific skills, you can refer to the pre- / post-questions for Chapters 5-18.

CHAPTER 5

Reviewing Addition

Questions to be answered:

Can the student:

1. Show how to use a number line.

2. Explain the principles of addition.

3. Use a number line and a calculator to add.

4. Solve addition word problems (using a structured form).

5. Use addition to mentally add small sums (or use calculator to add.)

Teaching Math, Book 1 has two chapters on addition and two chapters on subtraction. However, students who are hands-on learners need frequent experiences with addition and subtraction in order to be able to remember how to use these processes. The next two chapters will therefore review some of the principles taught in those four chapters and provide some examples that will help the teacher assess the student's understanding of addition and subtraction. Using addition in word problems will also be introduced.

The assumption of the two *Teaching Math* books is that students with Down syndrome and other hands-on learners will eventually be using a calculator for many, if not all, of their calculation needs as adults. However, we don't want them to just blindly input numbers into the calculator without some sense as to what is happening when they add, subtract, multiply, or divide. Volume 1 of *Teaching Math* details my rationale for using the calculator early and often with students who are concrete learners. If you have not read Volume 1, refer to Appendix A, Use of the Calculator, for tips on selecting a calculator and on teaching students a procedure to use in entering numbers into the calculator.

Reviewing Use of a Number Line

A number line is very useful in helping concrete learners understand what happens during addition and subtraction. Since the numbers are laid out in order from left to right, students can actually see how the numbers get larger as you move to the right when you add and smaller as you move to the left when you subtract. Some students understand more clearly if you make a vertical number line, with the numbers running up and down rather than right and left.

WHAT'S MY NUMBER GAME

OBJECTIVE: The student will illustrate that he can locate numbers on the number line and show the concepts of greater (bigger) than or less (smaller) than.

MATERIALS:
- Number line 1-100 from Appendix B (one per student), cut into 10 lines that can be attached together with tape or Velcro so that it forms a long line from 1 to 100
- Paperclips or tokens

SUCCESS STEP: Ask the student to assemble his number line on a table or on the floor in a long line from 1 to 50.

PROCEDURE:
1. The teacher tells the student, "I am thinking of a number between 1 and 50 and I want you to guess it. I will tell you whether the number you guess is too big or too small."
2. Allow the student to guess a number. Let's say your number is 24 and he guesses 10. Tell him, "Too small. Put the token (paperclip) on the 10 of the number line to help you remember that 10 is too small… Now, don't guess anything smaller than 10 because you know that 10 is too small."

3. If the student then guesses a number that is larger than your number, such as 30, tell him that number is too big, and tell him to put a token (paper clip) on the number line to mark which numbers are too big. Tell him he should now guess numbers between 10 and 30 and then have him move the paperclips on the number line as he narrows down the range. Pretty soon he will guess the right number.
4. Demonstrate the game one step at a time, having the students place the clips as you instruct them. If a student does not get the concept of guessing within the range, put pieces of paper over the right and left ends (or top and bottom) of the line, leaving only the numbers within the range showing. Set up a game of What's My Number with students and teacher taking turns thinking up a secret number.
5. If the students want to keep score, count how many trials it takes to guess a number. If you set up a game of teacher versus students, the students will probably enjoy the game more.
6. When the students understand the game, have them extend their number lines to 100. The game will take longer than when the line went only to 50.

ADDITION
Principles of Addition

1. Addition is used when two or more groups are put together.
2. The numbers can be added in any order (the commutative principle).
3. Two numbers can be added by *counting on* from the largest number.

Ideally, your student(s) will be familiar with the three principles above before proceeding further in this book. The student should have learned to visualize addition by manipulating straws, sticks, and other objects, by drawing pictures of the items, by making tallies, and finally by writing an abstract number sentence. You may want to review Chapters 13 and 14 in *Teaching Math, Book 1* if the student needs to practice visualizing addition in these ways. For this book, we will use a number line for visualization.

USING A NUMBER LINE AND CALCULATOR FOR ADDITION

OBJECTIVE: The student will be able to add with a calculator and illustrate work using a number line (for sums under 100).

MATERIALS:
- Number line 1-100 from Appendix B, cut into 10 lines that can be attached together with tape or Velcro to form a long line from 1 to 50 and then 1 to 100.
- Calculator
- Tokens, poker chips, or pennies
- Slate, whiteboard, or piece of paper
- Worksheet on Number Line Addition from Appendix B
- Pencils
- Addition Worksheet from Appendix B (optional)

SUCCESS STEP: Ask the student to take the 1-30 lines from the number line and put them in order.

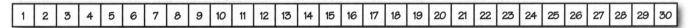

PROCEDURE:

1. Ask the student to point to the number 11 on the number line.
2. Have the student put a token on the number 11.
3. Show the student the slate with the problem 11 + 5 = ____.
4. Tell the student that you are going to add 5 to the 11 on the number line by *hopping* the token 5 more hops. Put your finger on the 11, count out loud and show him how you hop the token to 16 on the number line.
5. Have the students do the same thing on their number lines.
6. Make up problems (sums up to 30) to put on the slate and have the students work them on the number line.
7. Set up two more lines so the number line goes from 1-50. Make up more problems with sums under 50 and have the students do them together in the group.
8. When the students seem secure using the number line, assign the worksheet on Number Line Addition from Appendix B. Do not let the students use their calculators on this assignment.
9. At the next math session, have the students check their answers with their calculators. For extra practice, you can also have them do the Addition Worksheet with their calculators.
10. If the students are still having problems with addition, you can go to *Teaching Math, Book I,* Chapter 13, Beginning Whole Number Addition, and Chapter 14, Using Whole Number Addition for other teaching strategies.

Note: Be sure to make the observation verbally that the answer in addition is always larger than any of the numbers that are being added. It will help the students later when they are deciding what operation to use.

ADDITION ACTIVITY: USING DOUBLE DIGIT OR HIGHER NUMBERS

For students in secondary school and beyond, a meaningful activity may be to add up the calories for the things that they choose for lunch or dinner. They will get practice adding larger numbers on the calculator, and they will not have to worry about dollars and cents.

To work with my group of adult learners, I made a simplified list of many of the foods they ate for lunch along with approximate calorie counts. I required them to choose foods from several food groups. Besides getting practice adding larger numbers, they had continued exposure to the calorie counts of lunchtime or dinner favorites, helping them to learn about wise choices. They didn't have to reveal their choices to the other students.

In compiling my list of foods, I felt it was more important to list calorie counts students could easily remember rather than getting them precisely right; therefore, the counts are approximate. There are many calorie books that you can consult if you need calorie counts for additional foods. Some websites that may be helpful are:

- www.healthyweightforum.org/eng/calorie-counter
- www.calorie-counter.net
- www.caloriescount.org

You can also find calorie counts for most fast foods on the restaurants' websites (e.g., McDonald's).

If the students are buying school lunch, you should be able to find out the serving size and perhaps the calories from the lunch workers. At home you can get calorie counts from cans or food packages. Don't worry about being exact; you just want the students to have an idea of which foods are higher calorie and be able to practice adding together numbers that have some meaning for them.

For younger students, you might have them do their lunch calorie count for about two weeks. Older students could do either lunch or dinner calorie counts for as long as they are interested. If you have students who are counting carbs or doing Weight Watchers, you can use their numbers instead.

OBJECTIVE: The student will be able to add up 3-4 double (or more) digit numbers in a meaningful situation such as when planning or recording lunch or dinner menus.

MATERIALS:
- Calorie count books or calorie counts from restaurant websites (optional)
- Sample menu list with calorie counts (below)
- Blank Menu Plan form (Appendix)
- Paper and pencil
- Pictures of foods from ads in newspapers or magazines (4 pairs of pictures)
- Food packages or cans, if necessary

SUCCESS STEP: Ask the students what a calorie is. Do not expect an exact definition, but see if they relate calories to food. Then ask them what things have calories. You should get enough successful responses to start them off with praise.

PROCEDURE:
1. Show a picture of a fast food item and then a fruit. Ask the students which is the healthier choice. Repeat with other pairs such as a bag of French fries and a baked potato, a Whopper and a peach, a piece of cake or some lasagna. It will be more interesting to the students if you pick foods that you know they eat so the experience will be more real to them.
2. Kid with the students if they give the wrong answer and praise them for correct answers. Ask them why they think some foods are healthier than others. Agree with any reasonable answer.
3. Ask the students if they know *why* we really have to eat. Encourage the answer that we need energy from foods to run our bodies. Tell them that scientists have figured out a way to measure how much energy we get from a piece of food. The measure is called a *calorie*. Write the word on a piece of paper and show them how it is written.
4. Tell the students that the human body needs a certain number of calories a day to work properly, but if you get too many calories, your body can't use them all in one day. If that happens, your body stores the calories—as fat.

5. Ask the students how many calories we need each day. After their guesses, tell them that it depends on your age, size, and, as you get older, on whether you are a boy or a girl.

CHILDREN

Age	Calories needed per day
4-6 years	1800 calories
7-10 years	2000 calories

BOYS/MEN

Age	Calories needed per day
11-14 years	2500 calories
15-18 years	3000 calories
Over 18 years	2600 calories

GIRLS/WOMEN

Age	Calories needed per day
11-14 years	2200 calories
15-18 years	2200 calories
Over 18 years	2200 calories

Adapted from www.keepkidshealthy.com and wecan.nhlbi.nih.gov

6. Display the calorie chart. Have the students look at the table and point out how many calories they need for energy. Discuss mothers', fathers', and siblings' energy needs as shown by calories.

7. Ask the students how many meals a day they have—and snacks. Then discuss about how many calories we need at each meal. (You will probably have to divide the total calories needed per day by 3 for the answer.) Point out that many Americans eat more calories at their evening meal than at other meals and adjust the numbers to show that.

8. Show the students the calorie and food list and ask them to choose what they would like to have for lunch or dinner. Then add up the calories. For example, if I chose a lunch meat sandwich for 350 calories, an orange for 100 calories, milk for 100 calories, and a cherry pie at 400 calories, I would have a total of 950 calories for that meal.

9. Have the students compare that total to the amount recommended per meal for their age. Is it more or less than the total they should be aiming for?

10. Ask them what they have for snacks. Have them add up the total calories in the snacks they typically eat every day. See if it's almost as much as in a meal.

11. Check the items wanted for your meal on the Calories and Food chart. Write the calorie count for each selection in the Menu Plan and add to find the total calories.

GENERALIZATION ACTIVITIES:

Besides counting calories in meals, other activities can be used to practice addition:

- If the student uses a pedometer when walking or an odometer when riding his bike, add up the total number of steps or miles he walks a week or month. Or have him add up the number of laps he swims over a week or a month.
- At school (or home), set a goal to read 1000 (or another large number of) pages. Each time the student finishes a book, add it to a running total of the number of pages he has read.
- Play darts or ring toss and add up your score (in your head or with a calculator).
- Keep track of the number of points your favorite sports team (or player) makes over a period of time. Add up the points and compare with other students' or family member's favorite teams.

CALORIES AND FOOD

Food	Calories	✔
Main Dishes		
Hamburger	350	
Whopper	660	
Beef Burrito	400	
Chicken sandwich	530	
Roast beef sandwich	450	
Lunch meat sandwich	350	
Ham & cheese sandwich	560	
Pizza (2 pieces)	450	
Big Mac	570	
Desserts		
Cherry pie	400	
Shake	450	
Ice cream	350	
Frozen yogurt	180	
Sundae	750	
Candy bar	250	
Choc. chip cookie	100	
Cake	300	
Jello	100	
Diet Jello	90	
Side Dishes		
French fries	300	
Potato chips	300	
Bread (1 slice)	100	
Baked potato	100	
Mashed potatoes	150	
Rice	180	
Macaroni & cheese	280	
Baked beans	150	

Food	Calories	✔
Fruit/Vegetables		
Lettuce	10	
Green beans	10	
Mushrooms	10	
Tomato	10	
Corn	80	
Fruit cocktail (½ cup)	160	
Peaches (½ cup)	85	
Apple	130	
Banana	100	
Orange	50	
Strawberries (1 cup)	45	
Drink		
Diet pop	5	
Regular pop	160	
Milk	100	
Coffee/Tea	5	
Coffee/Tea (cream & sugar)	100	
Water	0	
Condiments		
1 T. butter	100	
1 T. Margarine	100	
Gravy (¼ cup)	100	
Mayonnaise	100	
Ketchup	16	
Mustard	16	
Relish	20	

MENU PLAN

Food	Calories
Main dish	
Fruit/Vegetables	
Dessert	
Side dishes	
Condiments	
Other	
Total	

Food	Calories
Main dish	
Fruit/Vegetables	
Dessert	
Side dishes	
Condiments	
Other	
Total	

Food	Calories
Main dish	
Fruit/Vegetables	
Dessert	
Side dishes	
Condiments	
Other	
Total	

A copy of this list with spaces for added foods is available in Appendix B.

Solving Addition Word Problems

Students need to be able to solve addition situations in daily life. In real life, addition problems are not usually phrased as neatly as they are in math books or in math class. However, learning to solve story problems in math class can help students learn to recognize situations that call for addition, as well as how to turn those situations into addition equations.

The language of story problems often poses difficulties for concrete learners. Many of these students use sentences that have simple syntax (grammar) such as "The boy (subject) ate (verb) the hot dog (object)." However, the question at the end of a story problem is a much more complicated sentence. The question at the end of the problem usually reverses subject and verb and begins with a question word such as *how* or a *wh* word. For example, "The Barbers drove 30 miles the first hour and 40 the second hour. How many miles did they drive altogether?" This type of sentence is called a transformation and is much more difficult for some students to understand than a simple declarative sentence with a blank left for the answer: "The Barbers traveled _____ number of miles in two hours."

To help your students understand the language of story problems, write the declarative sentence after the question sentence as often as necessary. Hopefully, if the students can learn to change the question at the end of a word problem to a declarative sentence themselves, they may be better able to understand the word problems they are given in school. However, they may need to have a lot of modeling to be able to write their own declarative answer sentence.

In addition, since the students are hands-on, concrete learners, use manipulatives, a number line, and drawing or acting out the problem to help make the solution visual. Try out several methods for making the story problems visual and let the students choose the methods that work for them.

Several published math programs (Swann, 2004; Semple, 2007) have students fill out a simple form to make solving word problems easier. I have included a similar form which will be expanded as students learn to do subtraction, multiplication, and division word problems. The form has to be used consistently if the students are going to learn to use it as part of their own problem solving.

Scott was quite competent at doing addition and subtraction problems with the calculator. However, when confronted with story problems, he seemed completely at sea and had no idea what process to use. His mother backed up and used yellow construction paper strips to represent the items in the problem. He began to manipulate the strips to see if he needed to add or subtract. However, he seemed to have little interest in the problems. His mother said, "Just pretend that these yellow strips are bananas, as in the story problem." He was silent for a minute, and then went to the kitchen counter and picked up two bunches of bananas. He showed much more interest in solving the problem using the real bananas. Even though he was competent with the calculator in addition, he needed to see the actual hands-on application to realize the value of addition.

CONCRETE OBJECTS TO REPRESENTATIONAL OBJECTS

As you work with your student(s) on solving story problems, you may find that some have difficulty moving from handling the actual objects to using more abstract problem-solving methods such as drawing tallies or circles to represent those objects or representing objects with numbers. We can make the steps a little smaller for them.

OBJECTS TO PICTURES

If you have a student who seems stuck on the stage of counting actual objects to add groups together, help him see that counting pictures of the objects gets you the same answer. For example, if the student is adding 5 balls plus 2 balls, you could match up each real ball with a realistic picture of a ball. Have the student count the real balls (7) and then count the pictures of the balls (7) and point out that those amounts are the same. Using pictures from ads or clip art, repeat the matching process with other problems. Say that John brought 4 apples for the fruit salad and Sarah brought 3 kiwi fruits. How many pieces of fruit do they have altogether? First do the problem with the real fruit. Then match each fruit with a picture and count the pictures. Compare that answer with the answer for the real fruit.

See if the student can do some simple addition using just pictures. If he is unable to do so, you might need to do more situations using real objects and pictures.

REPRESENTATIONS

Once the student can add pictures of objects rather than actual objects, you can progress to using more abstract representations. You could start out using the same fruit problem but give the student 4 single strips of paper to represent the apples and 3 single strips to represent the oranges. If the student doesn't understand what the strips stand for, write "apple" or "kiwi" on the strips or draw pictures of the fruit on the strips.

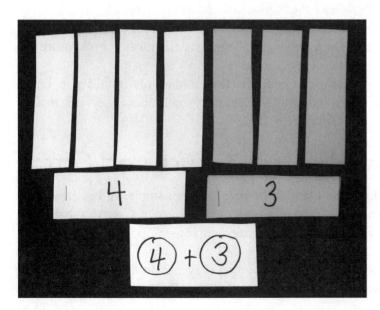

Lay the strips out in a line, with the 4 strips grouped together and the 3 strips grouped together. Show your student how you can add them together (count them all, or add on from 4 or use the calculator to add 4 + 3).

Next, staple the 4 strips together in one bundle and the 3 strips together in another bundle. Write 4 on the one bundle and 3 on the other. See if the student still realizes that you have to add 4 + 3 to get the total. If he does, move on to the next step, if not, have him count to see that there are still 4 and 3 strips in the bundles.

Next, show him that you can draw a circle to stand for apples and write 4 inside, and draw a circle to stand for kiwis and write 3 inside and put them together with a plus sign to find out how much you have in all. Write the numbers in the circles in the appropriate box on the form.

SOLVING PROBLEMS WITH HANDS-ON ACTIVITIES

OBJECTIVE: The student will be able to solve simple addition story problems by using manipulatives, a number line, drawing or acting out the situation, and/or writing a number sentence.

MATERIALS:
- Two volleyballs and 5 tennis balls or two other types of balls
- A 1-20 number line from previous activity and a token
- Paper and pencil or marker
- Addition form (in text and also in Appendix B)
- Addition word problems (in text and also in Appendix B)
- Calculator

SUCCESS STEP: Give the students a simple one-digit addition problem to solve. Praise their success.

PROCEDURE:
1. Introduce the following problem to the students by making the story visual:
   LaNell has two volleyballs and five tennis balls. How many balls does LaNell have in all?
2. Read the problem to the students. Have them write down the numbers involved. Model with them how you change the question at the end of the problems to an answer sentence with a blank. "LaNell has _____ balls."
3. Ask the students how they could draw, act, or show the problem on the number line.
   - Use actual balls to show how the balls can be counted together.
   - Draw two large balls and five smaller balls and count them for the total.
   - Put a pencil on the 5 of the number line. Then hop the token for two more jumps to land on 7.
   - Do the same story problem again, changing the numbers of the balls: e.g., Emily has one volleyball and four tennis balls. How many balls does Emily have in all?

Solving Addition Problems Using a Structured Form

Hand each student an addition word problem form (see samples A, B, C, and D on the following pages) and have him or her fill out the form just as in the hands-on experiences above. (Blank copies of the form are available in Appendix B.)

1. Students may balk at writing on the form when they can do the problem in their heads because of the small numbers in the problem. Tell them that we want to

learn the process with easy numbers so they will know how to do the more difficult problems. (They will need to know that addition makes the total larger than the numbers in the problem so they can tell the difference between addition and subtraction later on.)

2. In the acting-out stage, some students may need to use the actual objects (or pictures of them) at first. When the students are able to transition from using the actual objects to a semi-concrete representation, you can use straws or craft sticks to represent the items. I use construction paper strips about 2" long because I can write on each strip what it represents. For example, in transitioning from real bananas, I write the word "banana" on yellow paper strips.

3. Eventually, they can choose between drawing/acting out the problem or using a number line because some problems lend themselves to one kind of semi-concrete solution more than another.

4. At first, it is important that the teacher revise the question sentence to one that has regular word order; e.g., "The total number of marbles is ____." The student needs to see how this change can be made, but will be more apt to learn it from seeing many models beforehand.

5. Have the students do the following addition story problems with you, helping them fill in the form as they go along.

6. Have the student do the sample addition problems above on the forms. Start working out the problems orally as a group, then see if the students can do them independently.

7. Have the student do the Addition Word Problems Worksheet in Appendix B.

Chances are, your student or students will need more practice with addition word problems than is given here. You may want to use simple steps to help them succeed in learning:

- Go through the given problems with the students orally. Ask them what the problem is asking. Talk about whether your answer is going to be bigger than the biggest number in the problem, etc.
- Then have the students go through the *same* problems independently.
- Next use the same problem using different *names* than in the original problems.
- Then use the same problem using different *numbers* than in the original.
- Finally have them do new problems which are similar to the original problems.

The above suggestions were adapted from the method given in *Hot Math,* a program from the National Center on Accelerating Student Learning (CASL), Vanderbilt University by Lynn S. Fuchs and Douglas Fuchs (2002).

Note: Students may pick up that certain words signal one operation. My son told me that "together" and "altogether" usually mean addition. He is right, but many words are deceiving. The word "total" usually means addition, but not always. For example: "Mark had 8 candy bars. His brother took 3 of his candy bars. What was the total number of bars that Mark had left? The word "total" in that case means the final amount of candy bars after subtraction—not addition. If possible, have the students think through the problem to get the answer rather than rely only on certain key words.

SAMPLE ADDITION PROBLEM A:
Emily has two volleyballs and five tennis balls. How many balls does Emily have in all?

READ AND UNDERSTAND THE PROBLEM *(Write needed numbers with their labels; e.g., 4 desks)* 2 Volleyballs, 5 Tennis balls *(Write answer sentence with a blank)* Emily has _____ balls. 1	**DRAW (OR ACT OUT) THE PROBLEM** 00 OOOOO *or* Picture of student with balls *or* Picture of a number line *or* Act out the problem with balls 2
FIND OPERATION: Will the answer be larger than the biggest number in the problem? _____ 3	**CIRCLE OPERATION** Larger = addition 4
WRITE NUMBER SENTENCE (e.g., 8 + 4 = ___) 2 + 5 = _____ **SOLVE PROBLEM** (e.g., 8 + 4 = 12) 2 + 5 = 7 5	**WRITE ANSWER SENTENCE** *(Write the answer in the blank of the answer sentence in square 1)* Emily has 7 balls. 6

SAMPLE ADDITION PROBLEM B:
Washington High has 7 drum and bugle corps members who play the drums. They also have 9 members who play bugles. How many members do they have altogether?

READ AND UNDERSTAND THE PROBLEM

(Write needed numbers with their labels; e.g., 4 desks)

7 members,   9 members

(Write answer sentence with a blank)

Washington has _____ drum and bugle corps members.

1

DRAW (OR ACT OUT) THE PROBLEM

0000000   OOOOOOOOO

or

use strips of paper that have the words *drum* and *bugle* on them

2

FIND OPERATION:

Will the answer be larger than the biggest number in the problem? _____

(The whole band will have to have more total members than one of the groups.)

3

CIRCLE OPERATION

Larger = addition

4

WRITE NUMBER SENTENCE (e.g., 8 + 4 = ___)

7 + 9 = _____

SOLVE PROBLEM (e.g., 8 + 4 = 12)

7 + 9 = 16

5

WRITE ANSWER SENTENCE

(Write the answer in the blank of the answer sentence in square 1)

6

SAMPLE ADDITION PROBLEM C:

Rowan sold 8 pizzas for the school fundraiser. Mark sold 4 pizzas. How many pizzas did they sell altogether?

READ AND UNDERSTAND THE PROBLEM *(Write needed numbers with their labels; e.g., 4 desks)* 8 pizzas, 4 pizzas *(Write answer sentence with a blank)* The total number of pizzas sold was _____. 1	**DRAW, USE NUMBER LINE, OR ACT OUT THE PROBLEM** 2
FIND OPERATION: Will the answer be larger than the biggest number in the problem? _____ *(The total number of pizzas that Mark and Rowan sold has to be larger than what either one of them sold alone.)* 3	**CIRCLE OPERATION** Larger = addition 4
WRITE NUMBER SENTENCE (e.g., 8 + 4 = ___) 8 + 4 = _____ **SOLVE PROBLEM (e.g., 8 + 4 = 12)** 8 + 4 = 12 5	**WRITE ANSWER SENTENCE** *(Write the answer in the blank of the answer sentence in square 1)* 6

SAMPLE ADDITION PROBLEM D:
Katri bought school supplies. She bought paper for $1, folders for $4, and pencils for
$1. How much did she spend on school supplies?

READ AND UNDERSTAND THE PROBLEM	**DRAW, USE NUMBER LINE, OR ACT OUT THE PROBLEM**
(Write needed numbers with their labels; e.g., 4 desks)	
$1　　　$4　　　$1	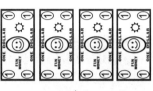
	$1　　　　　　$4　　　　　$1
(Write answer sentence with a blank)	
Katri spent $ _____ on school supplies.	
1	2
FIND OPERATION:	**CIRCLE OPERATION**
Will the answer be larger than the biggest number in the problem? _____	Larger = addition
3	4
WRITE NUMBER SENTENCE (e.g., 8 + 4 = ___)	**WRITE ANSWER SENTENCE**
	(Write the answer in the blank of the answer sentence in square 1)
$1 + $4 + $1 = _____	
SOLVE PROBLEM (e.g., 8 + 4 = 12)	
5	6

Commercial Games to Buy or Try

- Play the card game *Over and Out* (www.jaxgames.com).

    To play *Over and Out,* players add number cards to a central pile while trying not to exceed the number on a target card (22, 33, 44, 55, or 66). Each player's card either increases the total (by adding 1, 2, 3, 4, 5, or 10) or decreases it by 10 (subtract 10 card). Players who play a card that makes the total go over the target number lose 1 token, and are dealt 1 fewer card for the next hand, making it trickier to select cards to play that won't exceed the target number. This is a good game for mental addition of small sums. The cards requiring you to subtract 10 or multiply by 2 could be removed if you only wanted to practice addition.

- Play the game *Match 'Em* (www.jaxgames.com).

    One player puts out a card with a number on it. The next players have to match that person's number with an identical card or with cards that add up to the first person's number. If that is not possible, the next player draws two cards and puts out a new number to match. The person who knows his small value addition facts well has the advantage. A calculator can be used also.

- Play the game *Operation* (Milton Bradley).

    When a player successfully removes a part from the body, he receives a card saying how many hundreds of dollars he earned for the operation. At the end of the game, players add up their cards to find the total amount they have earned.

- Play the game *Math Dash* (Learning Resources).

    Players use number tiles to create equations using addition, subtraction, multiplication, or division on a Scrabble-like board.

CHAPTER 6

Reviewing Subtraction

Questions to be answered:

Can the student:

1. Explain the principles of subtraction.

2. Identify the types of word problems that use subtraction.

3. Solve all types of subtraction word problems.

4. Solve subtraction word problems (using structured form).

5. Solve mixed subtraction and addition word problems.

6. Write original addition and subtraction problems (optional).

Subtraction Using a Calculator

Volume 1 of *Teaching Math* details my rationale for using the calculator early and often with students who are concrete learners. If you have not read Volume 1, refer to Appendix A, Use of the Calculator, for tips on selecting a calculator and on teaching students a procedure to use in entering numbers into the calculator.

Principles of Subtraction

Subtraction is more difficult than addition because it is used in several different situations. The principles that need to be reviewed for subtraction are:

1. Subtraction is the reverse of addition. Subtraction can be checked with addition.

2. The order of the numbers is very important. The larger number is placed on the top in pencil-and-paper subtraction, and the larger number is put into the calculator first.
3. Subtraction is used when you want to know the difference between two numbers, and is called for in several types of problems:
   - "Take away" problems
   - Comparison problems
   - "How much more" problems
   - Change-back problems
4. The sign for subtraction is called minus and is a dash (—), as contrasted to the plus sign (+) that stands for addition.

Hands-on learners may have difficulty with the language used for subtraction problems. First, as was discussed in the addition chapter, many students with Down syndrome and other hands-on learners use sentences that have simple syntax (grammar) such as "The girl (subject) ate (verb) the ice cream cone (object)." The question at the end of a subtraction problem is usually in a form that reverses subject and verb and begins with a question word such as *how* or a *wh* word. For example, "The boys gathered 30 pieces of trash on the cleanup drive and the girls collected 24 pieces of trash. How many more pieces of trash did the boys collect than the girls?" The last sentence starts with a question word, then has the object followed by part of the verb, then the subject, and then another part of the verb. This type of sentence is called a transformation and is much more difficult for some students to understand than a simple declarative sentence with a blank spot for the answer. When students are first learning to do subtraction word problems, it will be helpful to rewrite the question sentence as a declarative sentence such as "The boys collected _____ more pieces of trash than the girls did."

Second, take-away subtraction is quite easy to visualize, but comparison, how-much-more, and how-much-change subtraction problems do not sound as if anything is being taken away. For example, "Paul had 4 marbles and Kyle had 9 marbles. How many more marbles does Kyle have?" Nothing is being taken away in that problem and the same is true of the other types of subtraction. Students just have to learn that those types of problems need the process of subtraction.

TAKE-AWAY SUBTRACTION

OBJECTIVE: The student will be able to use subtraction for take-away situations.

MATERIALS:
- Slate and chalk or whiteboard and marker
- 24+ cookies cut out of construction paper for demonstration (at least 12 paper cookies for each student) (Appendix B)
- Real cookies if desired
- Number line (1-30)
- Paper clip, penny, or other small token
- Take-away Subtraction Worksheet (Appendix B)

SUCCESS STEP: Hold out some paper cookies and ask the student to show you what the expression *take away* means. Encourage her to show you with actions.

PROCEDURE:

1. Tell the students that one of the students (use a student's name) had 24 cookies and another student came in and grabbed away 6 of them. How many are left? Act out the story with paper cookies.
2. Have the students count how many cookies are left (18).
3. Repeat the problem, telling each student to take away a different number of cookies. Then ask them how many cookies are left. You can laugh and make it a fun experience.
4. You may also do the problem using real cookies and having the students "take-away" 1 or 2 cookies at a time and counting to see how many are left.
5. Repeat the above procedure with other simple problems that you make up. Use concrete objects, pictures, or strips of paper in acting out the problems. It helps to personalize the problems by using the students' names.
6. Write the words "take away" on the slate or whiteboard so your students can visualize the name of the most common form of subtraction.

Using the number line

7. Now have the students subtract cookies using the number line. Have the student put her finger on the 24 on the number line and put a pencil there.
8. Tell the student you want her to take away 6. Ask the student to hop the token 6 steps backward and look what number is there. You have put a pencil on the 24 so the student can look back at the original number and visualize the "take away" action.
9. Repeat the above with several simple problems that you have personalized by using the students' names.
10. Continue practicing take-away problems (involving numbers under 30) until the students feel secure.
11. The students will probably need many concrete situations before they really understand take-away subtraction. When you feel they are secure, do the Take-away Subtraction Worksheet in Appendix B. To do these problems, the student crosses out pictures of the items that are taken away and then does the same problem written out.

COMPARISON SUBTRACTION

OBJECTIVE: The student will be able to demonstrate that comparison situations require subtraction.

MATERIALS:

- 12 paper cookies for each student and the teacher (Appendix B)
- Red construction paper squares and blue construction paper squares to be used as team labels
- Number line (1 – 50)
- Number Line Subtraction worksheet

SUCCESS STEP: Hold up some of the paper cookies and ask the students if they remember what they were used for in their last math class. Encourage them to say something about take-away subtraction. Praise correct answers.

PROCEDURE:

1. Hand out 6 construction paper cookies to one student and 3 cookies to another (or give the student 3 and yourself 6 if you are only working with one student). Ask the students who has more cookies.

2. Then ask, "How many more?" Have the students lay out their cookies in a line, matching cookie to cookie. They should be able to see that one student has 3 more cookies.

3. Tell them that they got the answer using subtraction. No cookies were taken away, but the same process is used for comparison.

4. Pair the students up, assigning half to the Red team and half to the Blue team. Give out the colored paper so each pair has one Red square and one Blue square.

5. Give each student 12 paper cookies.

6. Call out a comparison situation, saying, "Red team has 4 cookies. Blue team has 6 cookies. Put that many on the table in front of you."

7. Ask who has more cookies, Red or Blue. Let them all answer.

8. Tell them that you are going to *compare* the two amounts. Then ask them, "How many more?"

9. Let each student individually tell you how many more. Emphasize that they are using subtraction. Have them work out the same problem on their calculators using the minus sign.

10. Repeat the comparisons with every combination possible using 12 cookies.

Using the number line for comparisons

11. Pass out a number line (1 – 30) to each pair of students and tell them you want them to find how much more 6 is than 3. Using the number line, have the students place a pencil on the larger of the numbers (the 6). Have them put a token on the smallest number (the 3) on the number line.

12. Ask how they can find the difference between the two numbers. Hopefully they will tell you that they can count the numbers (or hops) between the two numbers they have marked to find the difference. Have the student demonstrate what to do. If no one is able to demonstrate, you show counting the hops and have the students do the same procedure on their number lines.

13. Give the students some more simple problems (e.g., 18 – 4) where the numbers are under 30. Have them demonstrate how to use a number line to find the answer while you are watching.

14. Continue with some comparison subtraction problems using the number line or acting the problems out. Try to come up with situations that the students will find meaningful. For example:

    - Kayla (use name of student) got 5 presents, but her little sister got 9. How many more did her sister get?
    - The students at (your school) get 8 days for Spring Break. The students at (rival school) get 12 days. How many more days off do they get?
    - Tyler is 13. His brother Nate is 16. How many years older is Nate than Tyler?

15. Then add the number line strips from 31—50 to make a line from 1-50. (Put it on a large table.) Have the students do some more simple problems with larger numbers (still under 50). Keep emphasizing the word *comparison* as you do each problem. Have the students chant the sentence "Comparison means subtraction!" as many times as seems appropriate.

16. When you are sure that they have the concept securely, have them do the Number Line Subtraction worksheet from Appendix B.

17. For the next session, have the students correct their worksheets using a calculator. Review with them how to put the problems in the calculator. Put the larger number first, then press the minus sign (—), then put in the smaller number and follow with the equals sign.

The Key Word Method

Some students rely on using the *key word method* for deciding which operation to use. For example, they learn to use subtraction if the problem uses *less than* or other similar words. Unfortunately, some question words or phrases can be deceiving as far as what operation is needed. For example, phrases such as "how much more"" or "how many more" used in comparison subtraction sound as if you should add because of the "more" and "many." This usage can be very confusing for students with language-related disabilities. The method used for story problems in this volume of *Teaching Math* emphasizes thinking through the problem to figure out the answer. Some of the students may pick up some of the reliable key words (such as "difference") on their own.

TOP THIS SUBTRACTION GAME

OBJECTIVE: The student will be able to subtract small value numbers mentally (or by calculator).

MATERIALS:
- Deck of playing cards with face cards removed and ace counting as 1
- Counters such as bingo chips, smooth stones, or buttons (50-100) in a box or bowl in the center of the table
- Calculator

PROCEDURE:

1. Shuffle the cards and divide them into two piles. Two students or the teacher and student play together in pairs.
2. Each student draws a card from her pile and shows it to her partner.
3. The player who has the largest number subtracts the other player's number from her number (may use calculator). If she gets the correct answer, she draws that number (the difference) of counters and puts them in front of her.
4. The game continues until all the cards have been drawn. The player with the most counters wins.

Uses of Subtraction in Word Problems

In addition to the two types of subtraction problems reviewed above, students may also encounter two other types of problems in real life:

1. "how much more" problems, and
2. "change-back" problems.

"HOW MUCH MORE" SUBTRACTION

When you know the price of an item, but you don't have enough money to buy it, you need to use "how much more" subtraction. You look at how much money you have now and figure out how much more money you need to save or earn to purchase that item. This problem essentially involves finding the *difference* between the two amounts, but students have difficulty understanding that it calls for subtraction. Since this situation commonly comes up in students' daily lives, I am considering it a separate category of subtraction to be taught.

> *For example:*
> Heather is saving to buy a CD. The CD costs $12, but she only has $6 saved up. How much more money does she need?
> (Heather needs $_____ more to buy the CD.)

*The student should learn that the amount of money she needs will **be smaller** than the price of the item.*

HOW MUCH CHANGE

Another very common situation that requires subtraction occurs in figuring what the change should be if you have given the clerk an amount larger than what is needed. The student needs to know that the change she gets back will always be smaller than what she gave the clerk—therefore, we use subtraction.

For example:
Mark is buying a pair of boots. The boots cost $70 including the tax.
Mark had four $20 bills ($80) to give the clerk.
How much change should he get back?
(The amount of change he will get back is _____.)

(See Chapters 13-15 for more information on teaching about change.)

The real difficulty comes in word problems when the students need to see these questions as requiring subtraction. Sometimes, learning to name the types will help the student know when to use subtraction. The following exercise focuses on helping students recognize the types of subtraction. However, if a student is not able to name the types, but does recognize that those types require subtraction, you do not need to insist that she learn to label the types of subtraction.

RECOGNIZING TYPES OF SUBTRACTION (OPTIONAL)

OBJECTIVE: The student will be able to recognize the types of subtraction and label them appropriately.

MATERIALS:
- Pencils
- Types of Subtraction Worksheet from Appendix B for each student

SUCCESS STEP: Draw a minus sign on the paper and ask the student what that sign means in math. Praise correct identification.

PROCEDURE:
1. Review with the students the different types of situations where we use subtraction—take away, comparison, how-much-more, how-much-change.
2. Read each word problem with the students and have them label orally the type of subtraction that is needed.
3. Some of the word problems on the worksheet will be identical to the problems that will be used later on in introducing subtraction using a structured form. The familiarity of the problems will help students learn the new task of using the form.
4. When all problems have been discussed, have the students do the worksheet independently.

Subtraction with Larger Numbers (Using the Calculator)

The students should have learned to do simple subtraction with their calculators in Book 1. However, they will need practice using the calculator to subtract larger numbers. Above all, they need to remember to put the larger number in their calculator *first*. Remind them that in addition, it doesn't matter in what order you enter the numbers, but in subtraction the order is very important—***Largest Number First.***

CALCULATOR ACTIVITY

OBJECTIVE: The student will be able to accurately subtract numbers with two or more digits with the calculator.

MATERIALS:
- Calculator
- Subtraction with Larger Numbers worksheet
- Pencil

SUCCESS STEP: Have the students do the problem 9 – 6 = ____ on their calculators. Help each to be successful.

PROCEDURE:
1. Go through two problems on the worksheet with the students.
2. Have the students finish the worksheet independently.
3. If needed, change some of the numbers on the worksheet and have the students complete the same problems under your supervision until you feel that the process is almost automatic.

Solving Subtraction Word Problems Using a Structured Form

The process for subtraction uses a form similar to the one introduced in the previous chapter for addition. The student reads the problem and makes the answer sentence using a blank. (At first the answer sentence will be given to the student.) Then she draws, acts out, or uses the number line to make the problem visual. She determines that the operation will be subtraction because the answer will be smaller than the biggest of the numbers given. Next, she writes the number sentence and solves the problem. She then puts the correct answer in the blank of the answer sentence.

When the students are secure with doing subtraction using the form, have them do the Subtraction Worksheet (Using Structured Form) from Appendix B. Use the blank Word Problem Form (Addition and Subtraction) from Appendix B for the students to work the problems.

Most likely, your students will need much more practice with subtraction word problems. You may want to use these simple steps to help them learn with success:
- Go through the given problems with the student orally.
- Then have the students go through the *same* problems independently.
- Next, use the same worksheet using different **names** than in the original problems.
- Then use the same worksheet using different **numbers** than in the original.
- Finally, have them do new problems that are similar to the original problems.

SAMPLE SUBTRACTION PROBLEM A:
Heather is saving to buy a CD. The CD costs $8 but she only has $5 saved up. How much more money does she need?

READ AND UNDERSTAND THE PROBLEM *(Write needed numbers with their labels; e.g., 4 desks or underline them in the problem)* (Write answer sentence with a blank) Heather needs $____ more to buy the CD. 1	**DRAW, USE NUMBER LINE, OR ACT OUT PROBLEM** Draw the number of dollars needed; Cross out number of dollars that Heather already has.  *(Slash represents dollars already earned.)* 2
FIND OPERATION: Will the answer be larger than the biggest number in the problem? Is the number of dollars that she needs to earn larger than the price of the item? ($8) (No, smaller) *(The idea is that the stuent doesn't exactly know what the answer number will be, but she does know that the answer will be larger than the biggest number—or not.)* 3	**CIRCLE OPERATION** Larger = addition <u>Smaller = subtraction</u> 4
WRITE NUMBER SENTENCE (e.g., 8 + 4 = ___) 8 – 5 = ____ **SOLVE PROBLEM** (e.g., 8 + 4 = 12) 8 – 5 = 3 5	**WRITE ANSWER SENTENCE** *(Write the answer in the blank of the answer sentence in square 1)* 6

SAMPLE SUBTRACTION PROBLEM B:
Mark wanted to buy a bike helmet. The helmet cost $24. Mark gave the cashier $30. How much change should he get back?

READ AND UNDERSTAND THE PROBLEM *(Write needed numbers with their labels; e.g., 4 desks or underline them in the problem)* (Write answer sentence with a blank) Mark will get $_____ in change. 1	**DRAW, USE NUMBER LINE, OR ACT OUT PROBLEM** 2
FIND OPERATION: Will the answer be larger than the biggest number in the problem? Will the amount of money you get back in change be larger than the amount of money you paid him? (No) 3	**CIRCLE OPERATION** Larger = addition Smaller = subtraction 4
WRITE NUMBER SENTENCE (e.g., 8 + 4 = ___) $30 – $24 = _____ **SOLVE PROBLEM** (e.g., 8 + 4 = 12) _____ – _____ = _____ 5	**WRITE ANSWER SENTENCE** *(Write the answer in the blank of the answer sentence in square 1)* 6

SAMPLE SUBTRACTION PROBLEM C:
Lexy brought 12 bottles of root beer to the Community Center for a party. Her friend Sandy brought 7 bottles of diet cola. Who brought the most drinks? How many more bottles did she bring?

READ AND UNDERSTAND THE PROBLEM *(Write needed numbers with their labels; e.g., 4 desks or underline them in the problem)* 12 bottles—Lexy 7 bottles—Sandy *(Write answer sentence with a blank)* One girl brought _____ more bottles than the other girl. 1	**DRAW, USE NUMBER LINE, OR ACT OUT PROBLEM** Draw 12 bottles. Then draw 7 bottles and compare them. 2
FIND OPERATION: Will the answer be larger than the biggest number in the problem? (No, the difference between both girls' bottles can't be more than the largest number of bottles brought.) 3	**CIRCLE OPERATION** Larger = addition Smaller = subtraction 4
WRITE NUMBER SENTENCE **SOLVE PROBLEM** (e.g., 8 + 4 = ___) _____ – _____ = _____ 5	**WRITE ANSWER SENTENCE** *(Write the answer in the blank of the answer sentence in square 1)* 6

SAMPLE SUBTRACTION PROBLEM D:
You have 7 packs of gum. Then you give 3 of them to your brother. How many packs of gum do you have left?

READ AND UNDERSTAND THE PROBLEM *(Write needed numbers with their labels; e.g., 4 desks or underline them in the problem)* (Write answer sentence with a blank) You have _____ packs of gum left. 1	**DRAW, USE NUMBER LINE, OR ACT OUT PROBLEM** 2
FIND OPERATION: Will the answer be larger than the biggest number in the problem? (Will you have more packs of gum left after you give 3 away?) 3	**CIRCLE OPERATION** Larger = addition Smaller = subtraction 4
WRITE NUMBER SENTENCE **SOLVE PROBLEM** (e.g., 8 + 4 = ___) _____ □ _____ = _____ 5	**WRITE ANSWER SENTENCE** *(Write the answer in the blank of the answer sentence in square 1)* 6

SAMPLE SUBTRACTION PROBLEM E:
Karryl's family had their picture taken. The photographer charged $12 for the pictures. Karryl gave him $20. How much change did she get back?

READ AND UNDERSTAND THE PROBLEM *(Write needed numbers with their labels; e.g., 4 desks or underline them in the problem)* (Write answer sentence with a blank) Karryl received $_____ for change. 1	**DRAW, USE NUMBER LINE, OR ACT OUT PROBLEM** 2
FIND OPERATION: Will the answer be larger than the biggest number in the problem? (Can Karryl get more money back than she paid the photographer?) 3	**CIRCLE OPERATION** Larger = addition Smaller = subtraction 4
WRITE NUMBER SENTENCE **SOLVE PROBLEM** (e.g., 8 + 4 = ___) _____ ☐ _____ = _____ 5	**WRITE ANSWER SENTENCE** *(Write the answer in the blank of the answer sentence in square 1)* 6

SAMPLE SUBTRACTION PROBLEM F:
MacKenzie has 11 Barbie dolls. On a trip to grandma's house, she brought along 6 of her Barbie dolls. How many dolls does she have left at home?

READ AND UNDERSTAND THE PROBLEM *(Write needed numbers with their labels; e.g., 4 desks or underline them in the problem)* 11 dolls, 6 dolls (Write answer sentence with a blank) MacKenzie has _____ dolls left at home. 1	**DRAW, USE NUMBER LINE, OR ACT OUT PROBLEM** 2
FIND OPERATION: Will the answer be larger than the biggest number in the problem? (Can MacKenzie have more dolls left at home than the total of dolls she has?) (No) 3	**CIRCLE OPERATION** Larger = addition Smaller = subtraction 4
WRITE NUMBER SENTENCE **SOLVE PROBLEM** (e.g., 8 + 4 = ___) _____ ☐ _____ = _____ 5	**WRITE ANSWER SENTENCE** *(Write the answer in the blank of the answer sentence in square 1)* 6

OPTIONAL: Go back to the addition and subtraction worksheets that the students have already done. Cross out the answer sentence that has the blank on each sheet. Have the students write the answer sentence with the blank themselves. (The total number of items is _____.) They don't have to do the problems themselves. You are just seeing if they have the idea of changing the question to a declarative sentence with a blank. This step is optional because students do not have to master it before they go on.

MIXED SUBTRACTION AND ADDITION WORD PROBLEMS

OBJECTIVE: The student will be able to distinguish between addition and subtraction problems and work accordingly.

MATERIALS:
- Mixed Addition and Subtraction Problem Worksheet from Appendix B
- Number lines (as used above)

SUCCESS STEP: Ask the student to show you 10 take away (minus) 4 on the number line. Model and have student repeat if done incorrectly.

PROCEDURE:
1. Have the students show you 8 – 2 on the number line.
2. Using the worksheet from Appendix B, have the students read aloud the problems one by one.
3. Ask the students to tell you whether the answer is going to be larger or smaller than the numbers in the problem. Talk about using addition if the answer is going to be larger than any of the numbers in the problem. If the answer is going to be smaller than the largest number in the problem, they should use subtraction.

> *For example:*
> At Christmas time the family counts 12 presents under the Christmas tree. Four of the presents are for Mom and Dad. How many are left for the rest of the family?

The answer has to be less than the largest number in the problem because the number of presents left can't be bigger than the total amount of presents. Therefore, the process must be subtraction.

4. Go through the entire worksheet with the students, not doing the problems, but having them decide on the operation. Keep a record of their choices. If they have trouble deciding, have them make a simple picture as was done on the form—for example, draw 12 squares for presents and then cross out 4 of them for Mom and Dad.
5. Then it is time for the students to actually work the problems with their calculators.

The second line on the form (shown on the next page) is the question "Is the answer going to be larger than the biggest number in the problem?"

Larger answer ➡ Addition
Smaller answer ➡ Subtraction

Written this way the form can be used for addition or subtraction. This method of determining the operation needed requires that the student think about what the answer will be ahead of time and requires that she read the problem carefully. This structured approach will be expanded with the processes of multiplication and division later.

READ AND UNDERSTAND THE PROBLEM *(Write needed numbers with their labels; e.g., 4 desks or underline them in the problem)* (Write answer sentence with a blank) 1	**DRAW, USE NUMBER LINE, OR ACT OUT PROBLEM** 2
FIND OPERATION: Will the answer be larger than the biggest number in the problem? ____ 3	**CIRCLE OPERATION** Larger = addition Smaller = subtraction 4
WRITE NUMBER SENTENCE **SOLVE PROBLEM** (e.g., 8 + 4 = ___) 5	**WRITE ANSWER SENTENCE** *(Write the answer in the blank of the answer sentence in square 1)* 6

THE JOURNEY GAME

OBJECTIVE: The student will be able to do simple addition and subtraction number and word problems in a game context.

MATERIALS:
- Die
- The Journey game—part of original informal assessment from Appendix B
- Problem cards for addition and subtraction—both number and word problems
- Game pawns

PROCEDURE:
1. Place the game pawns on Start.
2. One student throws the die. She moves her marker the number of dots she has thrown with the die.

3. The facilitator hands the appropriate card (addition number and word problems and subtraction number and word problems) to the student.
4. If the student can solve the problem, with or without the calculator, she moves the marker one place ahead if the problem is a number problem and two places if it is a word problem. (I usually let the students choose a number or word problem. However, if the students never choose word problems, I play the game with only word problems.)
5. The players must follow the instructions that are given on the spaces where they land. The player only has to obey a penalty square one time. If she lands on it again, she solves the problem as if it was a regular square.
6. The winner must land on the finish by exactly the right number. She may also win by doing a problem correctly and moving ahead one or two places.
7. You can control which problems the student gets or you can put the appropriate cards face down on the game board and let the student choose.
8. The Journey is a game that can be played over and over again. You can shape the practice by giving players number and word problems that fit what you want to review.

Note: When I was working with students to teach them how to do word problems with either addition or subtraction, they would sometimes ask me to quit teaching so they could play the Journey game. I then began using the problems on the Journey game cards to teach them. They quickly found out that if they paid attention while I was teaching, they would have the same or similar problems in the Journey game that followed and would be more likely to win.

Writing Original Word Problems (Optional)

Sometimes students can learn more thoroughly when they have to be the "teacher" and write their own problems. If the students are able to write three sentences easily, they may be able to do this on their own. If not, they may want to give the problems orally and see if the class can solve them. The rules for writing word problems are:

1. You must have a main character or characters.
2. You must have a problem to be solved.
3. You can use the processes of addition or subtraction.
4. You must have two or more sentences.

At first, you may give your student(s) a card with the number problem on it, such as $9 + 11 + 14 = $ _____. Then the student writes a story around those figures. For example: Three boys collected seashells at the beach. Allen found 9, Nathan found 11, and Ryan found 14. How many did they collect all together? The total number of shells is ____.

Have the students write their problems on a large index card in pencil. If necessary, have them dictate their word problem while someone else writes it down. The student should write the answer on the back of the card. You collect the cards and correct them with the student.

When the cards are ready, you can either present them to the class on an overhead projector or have students exchange cards with a partner.

Later you can have them make up the numbers also. You can control the size numbers that are used when you hand out the card with the numbers. At this stage, I suggest using numbers under 30. If the students are not able to come up with a scenario, suggest scenarios to them. For example, "There were 20 kids in the pool. Then ———."

As you can see, they will use language skills as well as math skills when devising their own word problems.

GENERALIZATION ACTIVITY—BUYING SNACKS

OBJECTIVE: The student will be able to do simple addition and subtraction problems in a simulated or real daily living situation.

MATERIALS:
- Menu board from Appendix B
- $1 bills for teacher (either play money or real)
- A $5 bill for each student

SUCCESS STEP: Ask the students what they like to eat when they go to a baseball game or an outside activity or picnic. Praise their contributions.

PROCEDURE:
1. Show the students the menu board and talk about the items and the prices.
2. Give each student $5 in play money to buy some snacks.
3. Ask each student to write down (or tell you) the two items she would like to "buy." Have her add to find the total for the two items. Then have her subtract that total from $5 to see how much money she would have left.
4. Go through the process of having each student pay for her snacks with her $5. Give her change and have her check to see whether you gave her the correct amount.
5. If possible, let the students choose two snacks from the menu and order them from a real snack bar at a movie theater, ball park, bowling alley, etc. Then they will really know why this skill is needed.

GENERALIZATION ACTIVITIES
- Use comparison subtraction when discussing sports scores. For example, Team A scored 77 points and Team B scored 65 points. How much did Team A win by?
- At home, tell your child she has 60 minutes to watch TV, do video games, be on the Internet, etc. After she uses 30 minutes (or another amount) subtract to see how much time she has left.
- When playing board games that use dice, have players subtract, rather than add, numbers to see how many spaces they can move. To make it more challenging, make your own dice using commercially available blank dice or the number cube template in Appendix B. Customize the dice with whatever numbers you want the student to practice subtracting with.
- Have your child count up how many calories she has eaten so far today (as described in Chapter 5). Then subtract from the total number of calories she should eat to see how many are left.
- Set individual or group goals such as to read 50 books, swim 100 laps, collect 300 cans for a food drive, walk 5000 steps (using a pedometer). Periodically subtract to see how many more you need to reach the goal.

CHAPTER 7

Introducing Multiplication

Questions to be answered:

Can the student:

1. Show, using concrete materials, that multiplication is repeated addition of like amounts.

2. Use the calculator (or mental math) to multiply single-digit numbers.

3. Demonstrate array multiplication (if needed).

4. Show with manipulatives and/or the calculator that numbers can be multiplied in any order (commutative principle).

5. Show with manipulatives that 0 times any number equals 0.

6. Multiply single-digit numbers by 10 without using the calculator.

7. Multiply single digits times multiple digits using the calculator.

8. Multiply multiple digits times each other using the calculator.

9. Decide whether an answer from the calculator is reasonable.

If a student understands the concepts of addition, multiplication can be introduced as a faster way to add numbers that are the same. Using the same colored plates and craft sticks that were used to teach addition in Volume 1 of *Teaching Math*, the student can experience how much more efficient it is to multiply numbers that are the same—even when using the calculator.

It is important that the student has a clear, vivid picture in his mind of what it looks like when you multiply numbers. He should see groups that have the same number of items, and he should understand why we count the number of groups and the number of items and then multiply. Activities where the students actually manipulate objects to find totals are the best way to anchor those pictures in their minds.

As with addition and subtraction, the question of memorizing the multiplication facts is a big factor in learning. Most people use a calculator for multiple-digit multiplication. Your student may be able memorize the single-digit facts with the help of rhymes, mnemonics, flashcards, and just plain hard work. See Chapter 8, Learning Multiplication Facts, for ideas on how to teach the facts. However, if you have persevered and the facts don't seem to stay with the student from day to day, teach him to conceptualize multiplication as described in this book and then allow him to use a calculator. Also, don't think that it must be an all or nothing situation with the facts. For instance, it can be very useful for a student to know the 2s, the 5s, and the 10s without having to resort to the calculator if he can't memorize the other facts.

Multiplication As Repeated Addition

INTRODUCING THE CONCEPT

OBJECTIVE: The student will be able to show, using concrete materials, that multiplication is repeated addition of like amounts.

MATERIALS:

- 5 or 6 brightly colored plastic dessert plates (same color)
- 40+ small craft sticks
- A calculator for each student
- Index cards and marker

SUCCESS STEP: Put out two plates and put 3 sticks on each. Ask the student to count the number of craft sticks and tell you the total.

PROCEDURE:

1. Using the same plates used in the success step, put out 4 plates with 3 sticks on each. Ask for the total number of craft sticks. Let the students count each stick on each plate and get the total. They may have to touch each stick to do the counting. (See if they can "spot" the smaller numbers without having to count each stick. Watch their faces to see if they are counting each stick separately.)

2. Put an index card to the right of each plate. Say, "How many sticks on this plate?" (3) Write the numeral 3 on the left edge of the index card. For each of the 4 plates, write 3 on the left edge of the index card and place it next to the plate. Looking at the line of 3's, say, "When we need to put numbers together, we use what operation?" (addition)

3. Say, "3 + (write the plus sign on the right side of the index card), 3 + 3 + 3 is our number sentence." Have the students punch in 3 + 3+ 3+ 3 on their calculators and then push the equals sign for the answer. The answer should be 12. Tell them that they can find the answer without counting by adding the numbers together. Point out that each of the numbers to be added is the same.

4. Then tell them that you know a shortcut way to get the total—a way that is faster than counting and faster than adding, which is called **multiplication.**

5. Say, "I have 4 plates here. What number of sticks do I have on each plate? **(3)** For my shortcut way, I put the number of plates **(4)** in the calculator, then put in the operation sign for multiplication **(X)** and then the number of sticks on each plate" **(3).** Put up a chalkboard or whiteboard where the students can see as you write **"4 plates"** and **"3 sticks** on each plate." Write the number sentence **4 × 3 = 12.**

6. Demonstrate how to multiply 4 × 3 on the calculator and compare the answer to the answer you got by counting or by repeated addition.

7. Repeat the above step using 3 plates and 2 sticks on each plate.

8. Emphasize that multiplying is faster than repeated addition and counting.

9. Then say, "Let's see what happens when the numbers get larger." Have the students repeat the process using 7 sticks per plate and 5 plates. Repeat with other larger numbers. The students should be able to see that larger numbers take more time when counting or doing repeated addition, but not when multiplying.

10. Each time, have the student tell you how many plates are in front of him, what operation to use (multiplication), and how many sticks are on each plate. Then put the numbers in the calculator and solve the problem. Sometimes count the total number of sticks before using the calculator, sometimes do it after you have the calculated answer.

11. When the students are quite proficient at multiplying the number of sticks using the calculator, ask one of them to set up a problem for you or for the other students using plates and sticks. This is one way that you can see whether they understand that the number of sticks on the plates has to be the same in order for them to multiply. If they do it wrong, ask, "OK, remember that to multiply you have to have groups of the same number. How many sticks do we have on this plate? And this one? Is that the same number?" Practice for 10-15 minutes each day until you feel that the students have the concept.

Season was quite proficient at addition. When she was asked to set up a multiplication problem with plates and sticks, she quickly put down the plates and sticks. However, when the other students tried to solve her problem, they were unable to multiply because she had a different number of sticks on each plate. Her instructor asked her to tell how many sticks she had set up. She quickly counted them correctly. The instructor told her that she did the addition problem correctly, but she had not done multiplication. The instructor then showed her what the multiplication problem would be by having her put the same number of sticks on each plate.

Multiplication of Single-Digit Numbers

INTRODUCING THE CONCEPT

OBJECTIVE: The student will be able to use the calculator (or mental math) to multiply single-digit numbers.

MATERIALS:
- Poker chips
- Colored plates
- Construction paper of 3 different colors
- White paper
- Calculators

SUCCESS STEP: Repeat a problem done successfully with the craft sticks and plates and have the student figure out the answer. Praise him or show him how to correct his error and give him an easier problem.

PROCEDURE:
1. Using the teaching method described above, put poker chips instead of sticks on the plates to see whether your student(s) can generalize the concept using different materials.
2. Have the students make problems out of the poker chips for you or the other students.
3. Tell the students that the plates are used to make it easier to see the *groups* that are on the table. The number of plates tells us the number of groups that we have. Ask the students what would happen if we put out pieces of construction paper instead of the plates. Would the answer of the number of poker chips be the same?
4. Put out 3 different colored pieces of construction paper (representing groups) to put the sticks or chips on. Use the word *groups* instead of plates while making multiplication problems using numbers 5 and under. Repeat with only white pieces of paper.
5. Fade out the pieces of paper and just put distinct groups of items in front of the student to introduce the use of the phrases *number of groups* and *numbers in each group* for multiplication.
6. With a variety of other kinds of plates and papers, have the students set up multiplication problems for each other. Then use straws or pencils or even chocolate chips so they can generalize multiplication to different objects. Use only numbers below the number 9. For example:
   - Two teams having 5 balls per team = how many balls in all?
   - Four groups of 3 different kinds of candy = how many pieces of candy?
   - Four plastic bags with 6 cookies in each bag = how many cookies in all?
   - Two rows of eggs having 6 eggs in each row = how many eggs in all?

Math Vocabulary

Heather understood the term multiplication, but she got mixed up on the word times as used in connection with multiplication. "I thought times was about the clock," she said. She needed to learn the difference between the several definitions of the word time.

Directly teach the vocabulary of multiplication, if needed.
- The student should know that the word *multiplication* means a quick way of doing addition with equal groups.
- He should understand that the symbol × signals multiplication and is read as *times.*
- If the student attends a general education class where learning correct terminology is important, teach the mathematical terms *factor* (one of the numbers being multiplied) and *product* (the answer in multiplication).
- Distinguish the symbol + for addition as contrasted with the × for multiplication.

In Book 1 of *Teaching Math*, Jan Semple's chant was used to teach the addition sign—saying, "Numbers together go plus, plus, plus," while striking the two index fingers together in straight lines showing a +. Likewise, you can teach the symbol for multiplication by using *arms crossed at the wrists* to show a slanting **x**. The manipulation

of fingers versus arms may help the students tell the difference between the addition and multiplication signs.

Students must be able to visually discriminate between the addition and multiplication symbols on the calculator. First try writing the two symbols on index cards with a marker. As you hold the symbols up, one by one have the students call out either "plus" or "times." Then reverse the activity and give the students the index cards. Now when you call out either *plus* or *times,* the students should hold up the correct index card. You can also play a second version of the game by using the words *multiply or add.* If you want to challenge some of the students, you can use both sets of words.

If the students are still having difficulty choosing the correct symbol, tell them that the addition sign is a child who stands with one leg on the ground (+). The multiplication sign is a grown-up who has two legs firmly on the ground (x). They learned about addition first (when a child) and multiplication later (when they were more grown-up).

Teach the students how to find the × sign on their calculators and go through each problem you model with real objects, watching the students' accuracy on the calculator.

GENERALIZATION ACTIVITIES

The students now need to see the value of being able to multiply in real life.

- Practice figuring out how many items total there are with things that come in groups. First count them and then multiply to get the answer. Suggestions are:
    - Several six packs of soda
    - Several bundles of juice boxes
    - Several packages of pencils
    - Packages of erasers
    - Small packages of peanut butter or cheese crackers
    - Packages of socks
    - Any other common items that are packaged with more than one item and can be viewed without opening the package (2 or more packages)

Have the students count the items in each of the packages and record and label the final number on paper (e.g., 6 socks). Then ask them if they would like to try a shorter way to find the total. Write the number of items in a package times the number of packages; for example:

3 (packages) × 2 (number of items in each packages) = 6

Figure out some situations in their everyday life where they *need* to use small figure multiplication:

- When it is appropriate, have a treat where each person gets 2 or 3 pieces. Tell the student to get the treats and distribute them. First, however, the student needs to figure out how many treats to get out. He needs to multiply the number of people times the number of treat items that each person is to have.

5 (people in family) × (times) 2 (the number of gummy bears
hat each person can have) = 10 gummy bears.

This activity can be repeated over and over when your child is responsible for handing out silverware or foods at home. In school the student could be responsible for handing out construction paper or other materials where a person needs more than one item.

- Talk about what happens when you are getting ready for a birthday or holiday party. You have to decide what you are going to serve for food or what items you are going to give the guests. In order to know how much food or other items to buy, you have to multiply the number of guests times the items or food each should have:

Number of balloons (3) × Number of guests (10)—
$$3 \times 10 = \underline{\quad\quad}$$

Number of miniature candies (5) × Number of guests (10)—
$$5 \times 10 = \underline{\quad\quad}$$

Number of ice cream sandwiches (2) × Number of guests (10)—
$$2 \times 10 = \underline{\quad\quad}$$

Moira looked at the problem 4 × 2 and wrote down the answer 8. The teacher asked her to draw a picture for 4 × 2. She drew four dots under the 4 and two dots under the 2. She drew 6 dots even though she had said the answer was 8. She really didn't know whether to add or multiply.

MULTIPLYING WITH FOOD

OBJECTIVE: The student will be able to transition from doing multiplication with concrete objects to doing semi-concrete or representational multiplication.

MATERIALS:
- Colored construction paper
- Marker
- Scissors
- Fruits such as grapes, strawberries, or bananas

SUCCESS STEP: Ask the student to identify the fruit. Praise success or help him be successful.

PROCEDURE:
1. Have the students do a simple problem with the actual fruits:
   Mom wanted to give each member of the family 2 pieces of fruit. There are 3 people in the family. How many pieces of fruit should Mom get out of the refrigerator?

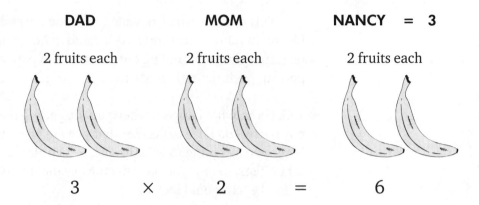

DAD	MOM	NANCY = 3
2 fruits each	2 fruits each	2 fruits each

$$3 \quad \times \quad 2 \quad = \quad 6$$

2. This problem should be easy to solve. The students can count the bananas.
3. Next, give each student 8 long strips of yellow paper (or blue paper if you are using grapes.)
4. Then tell the student that he should pretend that each strip is a banana. Write the word banana on a each strip, if necessary. Hand out a white piece of paper with the words Dad, Mom, Student's name on the top.

DAD   **MOM**   **Student's Name**

5. Ask the students how many bananas Dad was supposed to get (2). Repeat, asking about Mom and the student. Then have the students put 2 yellow pieces under the heading of Dad, 2 yellow pieces under the heading of Mom, and 2 yellow pieces under the student's name. Then write the equation on the white piece of paper:

$$3 \quad \times \quad 2 \quad = \quad 6$$

Have the students notice that the equations and the answer for the construction paper bananas are the same as for the real (or pictured) bananas.
6. Spend some time doing real (or pictured) small number problems and matching them up with construction paper strips.
7. Use the same matching type procedure with the construction paper strips and drawing dots on the paper. Show how this can be drawn as an array.

No. in each group	No. of groups ⬇		
	1	2	3
1	◯	◯	◯
2	◯	◯	◯

The student should draw 2 rows of 3 dots each. (It doesn't matter in what order the dots are drawn in multiplication.)

You can see that drawing out multiplication problems quickly gets tedious as the numbers get larger. Since my son has trouble with fine motor skills and drawing, I had him practice using straws bundled in groups of ten (and some singles). The numbers need to be small when working with manipulatives and multiplication.

CALCULATOR PRACTICE

Write the following problems as large as the student needs on unlined paper. Leave plenty of room between problems.

$2 \times 3 = \underline{\quad}$		$1 \times 2 = \underline{\quad}$
$3 \times 3 = \underline{\quad}$		$2 \times 2 = \underline{\quad}$
$2 \times 4 = \underline{\quad}$		$4 \times 3 = \underline{\quad}$
$4 \times 1 = \underline{\quad}$		$4 \times 4 = \underline{\quad}$
$5 \times 5 = \underline{\quad}$		$1 \times 5 = \underline{\quad}$
$5 \times 4 = \underline{\quad}$		$5 \times 3 = \underline{\quad}$

Have the student solve the problems with a calculator or by drawing dots. Check the answers by using the calculator and calling out the problem and the answer. Have the student check his own paper. Repeat as many times as necessary, mixing up the order of the problems.

Rowan was having trouble understanding the concept of multiplication. She was a strong tactile learner but a poor visual learner so her teacher figured out a way to show her tactilely. She had Rowan close her eyes. She took her finger and said, "Let's tap out groups of four with your finger." The teacher then tapped Rowan's finger on the table four times. She said, "That's one group of four." She then tapped out another group of four taps and said, "That is two groups of four." She repeated the motions with a third group of fours. Then she said, "Now we have tapped three groups of four. And we tapped 12 times."

Later the teacher had Rowan count on her fingers for each group that was tapped. She followed up with having Rowan combine 3 groups of four straws and then count them. After several tactile sessions, Rowan understood what multiplication was compared to addition.

MULTIPLY THIS GAME

OBJECTIVE: The student will be able to use the calculator (or mental math) to multiply single-digit numbers.

MATERIALS:
- Large dice (or blank dice) and a marker
  (Blank dice may be found at teacher's stores or on the Internet. You can also make dice from the number cube pattern given in Appendix B.)
- Index cards
- Opaque tape such as masking tape
- Calculators

PROCEDURES:
1. Cover the dots on the dice with tape. With marker, number the sides of each die from 0-5 You want the student to get a visual picture of the numerals. You may also make cardstock dice from the model in Appendix B.
2. Use any number of students. If the student is alone, the teacher will have to be his partner.
3. The first student throws both dice. The teacher then writes out a multiplication number sentence on an index card using the 2 numbers the student has thrown as the student reads the numbers on the dice. All the students then do the problem on their calculators, but only the first student is allowed to give the answer. If he is correct, the teacher writes the answer on the index card and gives it to the student. If the student gets the answer wrong, he does not get the card.
4. The next player throws the dice and play continues. If a student throws the side that says 0, he loses his turn. (This will introduce the fact that any number multiplied by 0 equals 0—which will be taught later.)
5. After about 15 minutes, call time. The winner is the person with the largest total when he adds up the totals on his index cards. For example, if Karl had the index cards $2 \times 3 = \underline{\textbf{6}}$, $5 \times 2 = \underline{\textbf{10}}$, and $3 \times 3 = \underline{\textbf{9}}$, for a total of $\underline{\textbf{25}}$, he would win over a student whose total was only $\underline{\textbf{20}}$.

Note: *Notice that the students must **add** up the numbers from each index card to get their own total. You can then see if they still understand multi-item addition.*

More Digits
6. Then introduce multiplying the numbers 6 through 9. For the game, you will have to write the numbers 6-9 on the dice, as well as 0. The students will be learning the more difficult facts from 6 to 9.
7. Encourage the students to memorize some of the facts as they are doing them on the calculator. They will find that it is faster if they memorize some of the easier facts rather than putting them in the calculator every time.

Array Multiplication

Sometimes students have a difficult time visualizing the multiplication of larger numbers. You can concretely demonstrate how to multiply larger numbers by using what is called *array multiplication*. *Array multiplication* is built on the principles of a hundreds board. Essentially, you make a chart that has the number of squares across the top of one of the factors (row) and the number of squares down (column) as the other factor. You then fill in the squares to find out what the total numbers of squares are.

1. Have the student count the books and then model the problem $8 \times 5 = 40$ on paper or chalkboard.
2. Tell the student that you don't have to draw all those books on every square. You can use dots or fill the squares with color.

	1	2	3	4	5	6	7	8	9	10
1										
2										
3										
4										
5										
6										
7										
8										
9										
10										

1. The array above shows $8 \times 7 = 56$. The student can get a visual picture of what multiplication does from this grid. He will also see that multiplication is a lot faster than counting all those squares. You can also have him add the number in each row, $8 + 8 + 8 + 8 + 8 + 8 + 8 = 56$ (repeated addition), and see that that, too, is slower than multiplication.

2. Have the student multiply 8×7 with his calculator. Again, he will see that although addition can get the right answer, multiplication is much faster.

3. Have the student show on the array grid on page 85 what $6 \times 7 = $ _____.

	1	2	3	4	5	6	7	8	9	10
1										
2										
3										
4										
5										
6										
7										
8										
9										
10										

For some students, it may be more helpful to show rows and columns of objects to help them visualize multiplication, as follows:

There are **four rows** of moons.

Each row has **six moons**

We have $6 + 6 + 6 + 6 = 24$

or $6 \times 4 = 24$.

SIXES GRID GAME

OBJECTIVE: The student will visualize multiplication with numbers from 1-6 by doing an array multiplication game.

MATERIALS:
- A Sixes Grid game for each student from Appendix B.
- Two dice (either with dots or with numbers 1-6, later 6-10)

PROCEDURE:
1. The first student rolls the dice. He multiplies the two numbers on the dice times each other (can use calculator). He locates the square where the two numbers intersect on the grid and writes the answer in that square.
2. The second student then throws the dice and fills in the answer on his grid.
3. Play continues until one person completely fills a line—across, down, or diagonally.
4. Later you may want to have the students cover the entire card to win.
5. Some students may have difficulty following the lines down and across. Two small rulers or a cardboard cut into a 90-degree angle may help the students find the intersecting square.
6. When you want to practice the facts from 6-10, you will have to tape over the dots on the dice, or use stickers. Then write the numbers 6-10, as well as 0 on the dice faces.

ORDER OF MULTIPLICATION

OBJECTIVE: The student will be able to show with manipulatives and/or the calculator that numbers can be multiplied in any order (commutative principle).

MATERIALS:
- Plates and sticks as used before
- Turnarounds worksheet (Appendix B)
- Calculators

SUCCESS STEP: Using a problem already mastered above (numbers 5 and under), have the student find the answer on his calculator. Praise his correct answer or give him an easier problem.

PROCEDURE:
1. Have the student set up a multiplication problem for you to solve on the calculator.
2. Ask him to do the following problems (which you have written on paper with plenty of space between them). Do the problems (*turnarounds*) 2 at a time.

$2 \times 3 = $ ____ $1 \times 2 = $ ____

$3 \times 3 = $ ____ $2 \times 2 = $ ____

$2 \times 4 = $ ____ $4 \times 3 = $ ____   and others, if needed.

3. Ask the student if he sees a pattern with these pairs. Lead him to notice that if you reverse the order of the two numbers, the answer is the same. Call them *turnarounds*.
4. Do a **chant** by crossing your arms just above the wrist to make an X, and saying, **"Multiplied numbers can *turnaround,"*** repeated three times.
5. Demonstrate with plates and sticks that you can switch the plates and sticks around and come up with the same answer. For instance, if you have 4 plates with 3 sticks on each one, you have the same number of sticks as you do with 3 plates with 4 sticks on each one.
6. Have the student do the Turnarounds worksheet from Appendix B.
7. Have the student show you some turnaround numbers with the plates and sticks until you are satisfied that he understands the concept.
8. With pairs of students, have one student write out problems as in step 2 above and then have the other partner write down the *turnaround* problem for that problem (at least 5 problems).

Multiplication by Zero

Katri consistently had trouble multiplying with more than one digit. She had learned her multiplication facts easily and did not need to use the calculator. After looking at quite a few examples of Katri's work, however, her teacher found that she did not understand that 0 times any number is 0. Whenever she encountered a 0 in a multiplication problem, she would add it rather than multiply by it. So she multiplied 4 times 50 and got 204.

$$\begin{array}{r} 5\,0 \\ \times\ 4 \\ \hline 204 \end{array}$$

Katri's teacher spent some time reviewing the facts with 0 in addition as contrasted with multiplication, and Katri didn't make that pattern of mistakes again.

Multiplying by 0 can often confuse students. They may confuse the problem with addition, where adding 0 to any number leaves the number the same. Acting out the situation may help them visualize the principle.

INTRODUCING THE CONCEPT

OBJECTIVE: The student can demonstrate with manipulatives that 0 times any number equals 0.

MATERIALS:
- 3 paper bags
- Small pieces of candy
- Small chalk or whiteboard or pad of paper with appropriate marker

● *Multiply This* game as done above
● Large dice (or blank dice) and a marker
   (Blank dice may be found at teacher's stores or on the Internet under "blank dice." You can also make dice from the model given in Appendix B.)
● Index cards
● Opaque tape such as masking tape
● Calculators

SUCCESS STEP: Put some candy inside one of the bags. Ask the student to look in a bag and tell you what he sees (candy).

PROCEDURE:

1. Put 2 pieces of candy in each of the 3 paper bags. Let the students look in the bags and figure out they have 3 groups of 2 each, or 6 pieces.
2. Next, take the candy out of bags. Let the students look in the bags. Ask the students how much candy is in the bags? (0) Show them on the board that $3 \times 0 = 0$.
3. Put your arms above your head with your arms rounded to make a zero. Chant the rule, "Zero times anything is zero." Repeat the Zero rule as much as is necessary to have them learn it.
4. When all have had a chance to try, give them some of the candy as a reward.

MULTIPLY THIS GAME

Repeat the **Multiply This** game played in the Multiplying by Single Digits section above, reinforcing the idea that players lose their turn when zero comes up on either of the dice. Have them chant the Zero rule, if appropriate.

Multiple-Digit Multiplication

I introduce multiple-digit multiplication by teaching students to multiply by 10. Multiplying by 10 by just adding a 0 to the end of the other number can be a useful skill and can help students realize what happens when multiplying by a two-digit multiplier. If possible, have the student discover the pattern himself.

MULTIPLYING BY TEN

OBJECTIVE: The student will be able to multiply single-digit numbers by 10 without using the calculator.

MATERIALS:
● Multiplying by Ten Worksheet (Appendix B)

- Straws or crafts sticks banded together as tens (50)
- Small chalk or whiteboard or pad of paper with appropriate marker
- Calculators for each student
- Cards with the numerals from 1-9 written one per card

SUCCESS STEP: Have the student put the problem 4×10 in the calculator and find the answer. Praise the correct answer.

PROCEDURE:
1. Show the problem $4 \times 10 =$ _____ on the board.
2. Say, "We are going to need groups of tens. How many *groups* of tens do we need?" Point to the number 4 on the board. "That's right—we need 4 *groups* of tens." Have a student bring you four groups of ten from the pile of 50 straws or sticks sitting on the table.
3. Ask the students if they can figure out how many straws or sticks are here all together. Some may think of skip counting by 10's to 40. Others may want to take off the rubber bands and count each group one by one. If no one figures out a way, you can demonstrate skip counting by 10's or counting each straw.
4. Write the answer (40) on the board in the appropriate place.
5. Write the problem $2 \times 10 =$ _____ on the board.
6. See if someone can figure out how to do the problem without much coaching from you.
7. Repeat with $3 \times 10 =$ ____, having the students do as much of the hands-on work as possible.
8. See if someone can figure out the pattern after the demonstrations. Do not tell the students what the pattern is until they have completed the Multiplying by Ten Worksheet and the following discussion.
9. Fold the worksheet so that only one problem at a time can be seen. Have the student do each problem one at a time with the calculator, unfolding the sheet as he goes. Check each problem to see if it is correct.
10. When the worksheet is finished, have the student look at each problem with its answer. Circle the important parts with highlighter, if necessary. Point out that the other number starts out in the ones or units place, but multiplying it by 10 moves that digit into the tens place.

$$5 \text{ (in ones place)} \times 10 = 50 \text{ (5 in the tens place)}.$$

See if the student can see that when you multiply 10 times a number, you just add a 0 to the right side of that number.
11. One by one, hold up the cards with the numerals 1 through 9. Have the students multiply each numeral by 10. Have then use their calculators to find the answers and show you the calculator answer window or write the answers using mental math on the back of their worksheets.

> *Kyle got the correct answer for the problem $3 \times 56 = 168$. However, he got the same answer for the problem $30 \times 56 = 168$. He had not learned the pattern for multiplying by higher multiples of 10. His mother had him use the calculator to do pairs of problems like 5×77 and 50×77 until he could demonstrate the pattern.*

SHOW ME GAME

OBJECTIVE: The student will be able to multiply any single digit by 10 without the calculator.

MATERIALS:
- Index cards with the numbers 10, 20, 30, 40, 50—one per card
  - Later 60, 70, 80, 90
- Sets of index cards with the numbers 1 through 5 on them (enough for each student to have a set)
  - Later 6 through 9
- Something that makes a pleasant noise to buzz for a wrong or late answer. You can just clap for the wrong or late answer, but the noise should not be harsh.
- Watch or clock

PROCEDURE:
1. Lay out the index cards with the multiples of ten in front of the student.
2. The teacher or another student randomly shows (one at a time) the index cards with the single digits on them. Firmly say the number that you are holding up. Say, "Touch the card that is 10 times my number."
3. The other student must touch the matching multiple of 10 card before the teacher softly counts 1...2...3. (Vary time according to students' strengths.)
4. If the student does not touch the multiple of 10 card in time or touches the wrong card, the teacher buzzes and he gets no points and his turn is over. If he is correct, the teacher or other student gives him the single-digit card that has been called. Then the other person repeats the sequence. They take turns for the whole game.
5. The game is finished when all the cards have been matched. The length of time the game lasts can be measured, and the student can try to beat his own time every time he plays. Some students will not do well when they are timed. If so, do not time them.
6. The student should then be able to explain the multiplying by 10 rule and demonstrate it with manipulatives or an array.

Single Digits Times Multiple Digits

The next step is to teach your student(s) to multiply the numbers 1-9 by two-digit numbers. Remind the students that they already know how to multiply one two-digit number—the number 10. Then use manipulatives to give them a tactile and visual picture of how that is done.

INTRODUCING THE CONCEPT

OBJECTIVE: The student will be able to multiply single digits by multiple digits using the calculator.

MATERIALS:
- Colored plates and sticks as used before (50 sticks and 4 plates for each student)
- Calculators for each student
- A slate and chalk or whiteboard and markers
- Calculator Multiplication Worksheets from Appendix B

SUCCESS STEP: Hold up 3 bundles of 10 sticks and ask one student how many sticks you have in your hand. If he is not correct, show him physically how you have 3 groups of 10 sticks each. Ask him how to write the same numbers on your board. You then write $3 \times 10 = 30$. Praise him for his help

PROCEDURE:
1. Ask the students if anyone remembers the shortcut way to multiply 10 times a number. Hopefully, they can tell you that you can multiply by 10 by adding a zero to the original number. Do several examples. If necessary, review the steps in the "Multiply by 10" section above.
2. Tell the students that they already know how to multiply by two digits because the number 10 has two digits.
3. Using the same example from the success step (3×10), ask the students what would happen if each of the 3 groups had 10 sticks and 1 stick more. Ask a student to add 1 stick to each of the 3 bundles of sticks. Have the students count the total. Encourage them to find a shortcut by saying they already know that 3 bundles of 10 equal 30 sticks and that they could just add the 3 extra sticks to make a total of 33.
4. Have another student make 3 bundles of 10 sticks and then add 2 more sticks to each group, making groups of 12. Write the numerals on the board: $3 \times 12 =$ ____. Count the manipulatives and fill in the answer (36).
5. Repeat step 4, having each student do the problems of 3×13, 14, and 15 using manipulatives. You can have the students work in pairs if you are working with a small group.
6. Repeat step 4 using 2 groups $\times$ 11, 12, 13, 14, and 15.
7. As needed, make up new problems and write them on the board, using numbers where the product does not exceed 50 (the number of sticks each student has). Use the manipulatives until you are sure the student has a visual picture of multiplying multiple digits.
8. Explain to the students that the calculator just multiplies the number of groups times the number of ones and then multiplies the number of groups times the number of tens and adds them both together to get the answer.
9. Show the students how this is done on the board. For example: $3 \times 14 = 3 \times 4$ and 3×10. $3 \times 4 = \underline{12}$ and $3 \times 10 = \underline{30}$. You then add 12 and 30 together to get the answer of 42. Have the students repeat your demonstration.
10. When the students clearly have a visual picture of what happens in multiplication, tell them that their calculator does the two multiplications and the addition for them and gives them the final answer. Have them do some of the preceding problems with their calculator and see the shortcut.
11. Have the students do the Calculator Multiplication Worksheets, using their calculators.

At this point, you may want the students to do some more problems from the mass market math workbooks that are sold in grocery and discount stores. The students should be able to use their calculators and write the answers in the workbook.

More Complex Multiplication

Sometimes we need to multiply by two or more digits in our daily life, especially when dealing with money. Most of us use calculators for accuracy in these calculations. The difficulty with paper and pencil calculations of multiple digits seems to be in lining up the partial products with the correct place value, remembering the carried or regrouped factor, and doing the final addition.

If it is necessary for the student to do paper and pencil calculations, I would turn the lined paper sideways and have him write the numbers in those columns to help him keep the ones and tens lined up.

The non-mathematicians among us are spared these difficulties by using the calculator. Once the student can multiply two-digit numbers by one-digit numbers meaningfully using a calculator, we can introduce more complex multiplication problems using the calculator without much more instruction.

We need to be sure the student is putting the problems into the calculator correctly, without undue speed, and check for accuracy frequently. Now a printing calculator may be useful, so the student can look back at the paper tape and see if he entered the numbers correctly. Or you may want to invest in a calculator such as the See 'N' Solve Visual Calculator or the TI-15 calculator (Texas Instruments) that displays the entire problem, including the answer, on the screen.

In addition, we need the student to look at the answer he got on the calculator and see if it is reasonable. General mathematics textbooks often ask the student to estimate the answer to see if it is reasonable. It is my opinion that estimating the answer can be more difficult for a student who learns in a hands-on manner than actually doing the problem. I have the most luck in getting students to look at the reasonableness of the answers when they have a real-life situation or a well-understood story problem. For example:

> *"Jake, do you really think you are going to need $219 for bus fare? That's an awful lot of money. I could probably fly to Chicago for that."*

> *"Do you think that I could bring in only two marbles and yet have enough marbles to give one to everyone in the class?"*

Have the students learn to ask themselves if the answer seems reasonable when they are able to see a hands-on or real-life situation. (Chapter 9 goes into more detail about using multiplication in real-life situations.)

See if the students can apply the principles learned in single-digit multiplication to multiplying multiple digits. Have them explain in words, if possible, how they did the problems with their calculators. If their understanding is secure, it is not necessary to have them do many multiple-digit worksheets just for practice, but a few problems introduced at the beginning of each lesson should help them maintain that skill.

BATTER UP MULTIPLICATION GAME

OBJECTIVE: The student will be able to multiply using the calculator in a game situation.

MATERIALS:
- Cards with 2-3 challenges (math problems)
- Cards with game calls
- Baseball game board (Appendix B)
- Game markers/tokens
- Calculators

PROCEDURE:
1. Put players' markers on home plate.
2. First player picks a card from the pile (which has been shuffled).
3. If it is a game call card (one with instructions), he follows the directions, and the play goes to the next person.
4. If he chooses a challenge card (one with problems), he can choose either of the two problems.
5. He solves the problem, with or without using the calculator.
6. He or the teacher finds the answer on the answer key.
7. If the player is correct, he may advance one base. If the player is not correct, he strikes out and his turn is over. Then the play goes to the next player.
8. The object is to keep answering the cards right so you advance around the bases and make a run. Chance, from the game call cards, as well as accuracy in solving the problems, makes a winner. The person with the most runs at the end of play wins the game.
9. The game can also be played with teams, alternating the turns between teams.
10. The game also has a score board that must be marked as the game progresses.

GENERALIZATION ACTIVITIES
- Involve your child in multiplication scenarios related to buying or preparing food.
  - Get your child to help you figure out what you need to buy for school lunches for a week. If you pack 3 juice boxes a day, 5 days a week, how many do you need for the week? If you use 4 slices of bread a day, 5 days a week, how many slices do you need?
  - Make s'mores. Everyone needs 2 graham crackers, 3 marshmallows, 4 squares of chocolate. How many of each do you need if there are 2 or 3 or 4 people?
- When shopping, point out products that come in multiples—for example, Pop Tarts come in 3 packs of 2 ($3 \times 2 = 6$), a box of packaged peanut butter crackers might contain 6 or 8 packs of 6 crackers, or 6 or 8×6, etc.
- If your student is learning about calories per serving, look at packages and determine things such as: if one cookie has 90 calories, how many calories will you get if you eat 2 or 3 cookies? If one serving of juice has 110 calories, how many calories will you consume if you drink 2 or 3 servings in a day?

- If your student needs batteries for his Christmas or birthday presents, talk him through which packs of batteries to buy at the store—2 packs of 6? 2 packs of 8?
- At school, talk about how tests or homework were graded, if numerical scores are given. For instance, each question was worth 3 points, so if you got 5 right, you got 15 points.
- When you are watching football on TV or at a game, talk about how many touchdowns (plus field goals) it takes to make 14, 21, 28, 35, etc. points.

Chapter 8 will give some hints on how to learn the multiplication facts. Chapter 9 describes how to use multiplication in real situations and in story problems. It is in this chapter that we can see whether the process of multiplication is useful for the students.

Learning Multiplication Facts

Questions to be answered:

Can the student:

1. Skip count by 2s, 5s, and 10s to 100 fluently and relate the skip counting to multiplying by 2, 5, and 10.

2. Explain the 1 and the 0 multiplication facts.

3. Say the 9 times tables using finger cues or the subtraction pattern.

4. Find facts on a 10 x 10 multiplication table.

5. Recite the 2, 3, 4, and 5 multiplication facts fluently.

6. Recite the 6, 7, 8, and 9 multiplication facts fluently.

Although it may be difficult for students with Down syndrome and other hands-on learners to learn all of the multiplication facts, I would try to teach some of the facts, especially those useful for handling money. If the students have worked in *Teaching Math, Book 1* (Basic Survival Math), they will be able to skip count by 2s, 5s, and 10s. Skip counting can easily be turned into the knowledge of the 2, 5, and 10 times tables.

Understanding the patterns of multiplying by 1 and by 0 will give them two more sets of times tables. Students can be taught that 1 times any number equals that number, and that 0 times any number equals 0.

I will discuss some of the ways that help any student to learn the multiplication facts. Remember, though, that if your student becomes overly frustrated in trying to learn the facts or frequently forgets them, you should switch your focus to helping her understand multiplication with the use of the calculator.

Multiplying by 2, 5, and 10

See if the students can count by 2s, 5s, and 10s by rote. If they miss many of the numbers, you may want to re-teach Chapter 9 in the first *Teaching Math* book. Review and expand on the skip counting with the following procedure.

REVIEWING SKIP COUNTING BY 2s

OBJECTIVE: The student will skip count by 2s and relate the counting to multiplying by 2.

MATERIALS:
- Number line that can be extended to 100 from Appendix B (The same number line used in Chapter 5 and 6)
- A penny or game marker
- A slate or whiteboard and chalk or marker
- Flashcards/index cards numbered from 2-50 by 2s

SUCCESS STEP: Ask the student to jump the penny or marker from 2 to 4 to 6 to 8 to 10 on the number line. Praise success or help the student jump the penny on the number line correctly.

PROCEDURE:
1. Ask what kinds of things come in 2s (twins, eyes, ears, hands, feet, arms, hands, shoes, gloves, halves of a bun). As the students name things in twos, try to draw them on the board or paper.
2. If the students don't remember how to count by 2s, teach them a chant:

   2, 4, 6, 8,10
   Let do it over again.

   12, 14, 16, 18, /20
   Don't know why we have so many.

   22, 24, 26, 28, 30
   That's enough—let's not get wordy.

3. Using the slate or whiteboard, write the numbers 0–9 in one row, the numbers 10–19 directly below that, the numbers 20–29 below that, and the numbers 30–39 below that. Point out the pattern of the tens number going up by one while the ones number repeats 2, 4, 6, 8, 0.
4. Make up flashcards with the twos and mix them up. Have the students put the twos in order (2, 4, 6, 8, etc.).
5. Practice counting by twos until the student can say them fluently. Use the chant frequently. If the student needs more visual cues, use the laminated number line and mark the two's with a washable marker.

6. Ask the student to skip count by 2s from 0 to 10 on the number line as you count the number of jumps aloud (1, 2, 3, 4, 5). Explain that counting by 2s is like multiplying the number of jumps by 2. Write on a board or paper:

$$1 \times 2 = 2$$
$$2 \times 2 = 4$$
$$3 \times 2 = 6$$
$$4 \times 2 = 8$$
$$5 \times 2 = 10$$

Point out the jumps on the number line and read the number sentences (equations) above.

7. Repeat the procedure above with skipping by 2s from 10 to 30.

8. Then ask what would happen if 3 people with 2 hands each came into the room? How many hands would be in the room? Draw a picture to illustrate with three rows of two hands each and the equation $3 \times 2 = 6$.

9. Repeat the drawings with some of the groups of items the students said came in 2s (twins, eyes, etc.)

10. After the students know the chant, have them touch each finger as they count 2s up to 10 (1, 2, 3, 4, 5). Tell them that the number of the finger tells them the number to multiply by. For example, to multiply 2×3, they would start at their first finger and say 2, touch their second finger and say 4, and touch their third finger and say 6.

11. Another strategy that may help students remember their 2s is to relate them to doubles facts in addition. Point out, for instance, that 2×4 is another way of saying you have two 4s (4 + 4). Refer back to pages 170-71 in *Teaching Math, Book 1* for some mnemonics and activities useful in teaching doubles facts.

REVIEWING SKIP COUNTING BY 10s

MATERIALS:
- Whiteboard or slate and markers or chalk
- Number line (0-100) from Appendix B
- Index cards (numbered from 10 to 100, by 10s)
- Straws or craft sticks in bundles of 10 (at least 10 bundles)

PROCEDURE:
1. Ask the students if they know the trick for counting by 10s. Remind them that they have already learned to multiply by 10 (by adding a zero to the counting numbers to skip count by 10's). Show them on the whiteboard or chalkboard: $1 + 0 = 10$, $2 + 0 = 20, 3 + 0 = 30$, etc.

2. Using their number line, show how you skip count by 10s.

3. Make index cards for the 10's facts up to 100. Mix them up and let the students put them in order. They usually learn the 10s quickly.

4. You can also use manipulatives (straws or craft sticks) that are bundled in 10's and have the students count the bundles—0, 20, 30, etc. Relate the number of bundles to multiplying by 10. Say, "$2 \times 10 = 20$. Pick up 20 straws." The student practices picking up bundles of 10 in answer to a multiplication fact. Have the students practice until they are fluent with the facts.

REVIEWING SKIP COUNTING BY 5s

Skip counting by 5s is especially important in counting money and time. Nickels and quarters can be counted by using 5s, as can five-dollar bills. Telling time on an analog clock may also require counting by 5s.

MATERIALS:
- Number line (0 – 50)
- Whiteboard or slate, markers or chalk
- Flip chart from Appendix B

PROCEDURE:
1. Using the number line extended to 50, you jump the penny or marker by 5s to 50. Explain that now that the students know the 10's, you are stopping halfway in between at the 5s.
2. Ask one student to count by 5s on the number line, as you have done. Write on the slate or board the numbers 5, 10, 15, 20, etc.
3. Point out that every other number ends in 5 and the in-between numbers end in 0.
4. You can make a small lift-up flap chart so the students can practice counting by 5s. See note on how to make flap charts below.
5. Have the students use their fingers to practice multiplying by five. You say, "4 × 5." They lift up four fingers and skip count to the fourth finger.
6. Repeat as needed. Have a pair of the students practice together, if possible.

Making a Flap Chart

To make a small flap chart that your student can use to teach and test herself, fold an 8½" x 11" piece of cardstock paper in half width-wise (5½" by 8½"). Cut the front half into 5 equal strips. Do not cut the back half. On the front strips, write 2, 3, 4, 5, and 6. On the corresponding back half, write 10, 15, 20, 25, and 30. The students can teach/test themselves by looking at the front strips and saying the 5s. They can check their answers by lifting up the front flaps to see the correct answer. Staple the top fold to make the strips easier to lift. Other flap charts can be made for the 5s up to 100. The front strips represent one multiplier and the back numbers are products of the 5s. (See photo of a flap chart in Chapter 12, page 176.)

Nine Times Tables

There are a lot of interesting things about the 9 times tables. Two methods for remembering them will be discussed here, but if you know a different way, feel free to try it with your student(s).

FINGER CUE METHOD

OBJECTIVE: The student will be able to say the 9 times tables using finger cues.

MATERIALS:
- A piece of paper for a drawing of the student's hands with the fingers numbered.
- A marker or crayon
- A slate or whiteboard and chalk or marker
- Multiplication Tables—6s to 9s (Appendix B)

SUCCESS STEP: Ask the student to trace around one hand on the paper. If this is difficult for her to do, ask her to wiggle each finger as you touch and count the fingers on her hand. Praise her attempt.

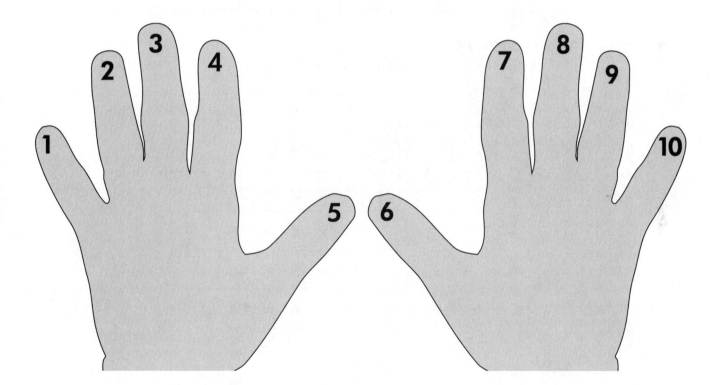

PROCEDURE:
1. Trace around each student's hands and label fingers on the drawing of each hand as shown in the picture above. Some students will be able to label the fingers themselves.
2. Write the problem 3 × 9 on the board.
3. Tell the students to multiply 3 times 9, they should fold their third finger down. If this is too hard for them to do, tell them to hold both hands slightly over a table or their thighs and then just touch the third finger down, leaving the other fingers straight. Looking at both of their hands, the first number (the tens) is represented by

the 2 fingers at the left of their folded-down finger. The second number (the ones) is represented by the 7 fingers to the right of the folded down finger. Therefore, $3 \times 9 = 27$.

4. Repeat the above with the rest of the nines facts.
5. Make sure students understand that this finger calculation method only works for the nines.

NINES METHOD

OBJECTIVE: The student will be able to say the 9 times tables using the nines pattern.

MATERIALS:
- A Times Nine chart from Appendix B

SUCCESS STEP: Have the student point to the multiples of 9 on the 9 chart (from the Appendix) as you name them out loud. Praise correct pointing.

PROCEDURE:
1. Give a copy of the 9's chart to each student.

9	× 1	=	9
9	× 2	=	18
9	× 3	=	27
9	× 4	=	36
9	× 5	=	45
9	× 6	=	54
9	× 7	=	63
9	× 8	=	72
9	× 9	=	81

2. Ask the students to name and touch each of the numbers in the SECOND column (2-9).
3. Then tell them to look and touch the first digit (tens) of the answers in the THIRD column.
4. See if they can figure out the pattern that the first digit in the THIRD (answer) column is one less than the digit in the SECOND column.

5. Go through each multiple of 9, getting them to figure out the first digit of the answer.
6. For the student to figure out the second digit of the answer, she will need to know what number added to the first number equals 9. The digits in each product (answer) add up to 9. For example:

$$2 \times 9 = 18, \ 1 + 8 = 9.$$
$$3 \times 9 = 27, \ 2 + 7 = 9.$$

If the student cannot yet do this, but is not able to manipulate her fingers to do the finger cue method above, you may want her to practice the addition facts that add up to 9 in order to use this nines solution.

7. Either of the above methods can be illustrated quickly but need to be practiced frequently until the student can do them automatically.

Other Facts

MULTIPLYING BY 1 AND BY 0

OBJECTIVE: The student will explain the pattern of multiplying by 1 and by 0.

MATERIALS:
- Chalkboard and chalk or whiteboard and marker

SUCCESS STEP: Ask the student to tell you the answer to one of the multiplication facts that you know that she knows.

PROCEDURE:
1. Ask the student what 1 times 2 is. If she is incorrect, tell her the answer is 2 and write $1 \times 2 = 2$.
2. Continue showing the student the 1 times table on the board.
3. Frequently practice the phrase, "1 times any number equals that number" with the students. Occasionally ask a student, "What is 1 times 3, etc.?"
4. Use the same procedure to teach that 0 times any number is 0. Repeat the demonstration with bags full of nothing from the previous chapter, if necessary, to help the students understand the concept.

MULTIPLYING BY 2, 3, AND 4

OBJECTIVE: The student will be able to say the 2, 3, and 4 times tables.

MATERIALS:
- 2s and 3s Multiplication Fact Cards (Appendix B)
- 4s Multiplication Fact Cards (Appendix B)
- Number line (1 – 50)
- Token

SUCCESS STEP: Review with the student some facts that she can recite easily.

PROCEDURE:
1. Using the procedures given above for skip counting, hop the token by 3s on the number line.
2. Cut out the 2s (which the student should already know) and the 3s Multiplication Fact cards. The answers are on a separate card.
3. Have the students work with 4 cards at first (2 problems and 2 answers) and match the problems to the answers.
4. Increase the number of fact cards used making sure that the student is successful.
5. Once the student is secure with the 2s and 3s, introduce the 4s as above, using first the number line and then the 4s Multiplication Fact Cards.

Teaching the Remaining Facts

COMMERCIAL
PRODUCTS
THAT MAY HELP

Most commercial texts suggest using flashcards for teaching the rest of the multiplication facts. They recommend first testing the student with the flashcards and then having her practice the facts that she misses. Teachers often play games using the facts so the answers will come almost automatically.

There are some alternatives to flashcards that may be more appealing for students with Down syndrome or other hands-on learners. One commercial product called *Rhymes 'n' Times* (www.rhymesntimes.com) uses mnemonics, rhymes, motions, and fun objects to capture the students' attention and help them learn. For example, the student is taught the chant:

3 × 4,
12 knocks on the door

She pretends to knock on the door as she chants and then goes to a real door and knocks 12 times. In another example, the chant is:

3 × 8
24 horseflies on my plate.

The student actually takes 24 plastic horseflies and puts them on her plate, counting them one by one. There is a fairly extensive kit with many objects that the students may find interesting, such as 42 rings, 49 bugs, 64 colored paper clips, and 28 spiders. The chants, rhymes, physical actions, and interesting objects help with short-term, working, and long-term memory. If you are really eager to have the student memorize the times facts, this book and materials might be worth the slightly over $100 cost. Make sure you keep track of all those objects and only use them for math purposes.

Another commercial program that uses mnemonics to help with fact learning is called *Memory Joggers* (www.memoryjoggers.com). In this system, each number always represents a particular person, animal, or object. For example, 8 is always represented as Nate, the snowman (and is drawn to look like a snowman), and 9 is always represented as Nina, the porcupine (and is drawn to look like a porcupine).

You get a set of large cards for each multiplication fact, together with a story that is meant to help you remember the fact. For example, for 8 × 9, there is a picture of Nate (8) together with Nina (9). Nina is shown crying and reading a letter. "Sent to the zoo, sent to the zoo" (which sounds like 72), she is wailing. The story goes on to explain that she got this letter and thinks she's going to be sent to the zoo, but then Nate reads the letter and explains that they just want her to send money to the zoo. The basic level of Memory Joggers costs about $40.

The website www.multiplication.com uses a similar picture-cue method to help a sticky working memory. The method, which can be previewed on the website for free, uses rhyming words, pictures, and stories for the more difficult facts. For example, for 4 × 5, there is a picture of a door (for 4), a beehive (for 5), and honey (for 20) dripping out of the hive. Door and hive can remind the student that when the door (4) of the hive (5) is left open, out drips the honey (20). Many students can remember the simple pictures that are shown, as well as the story that goes with it. The accompanying book, *Memorize in Minutes: The Times Tables,* has lesson plans, activities, worksheets, flashcards, tests, and charts (about $20-$25).

Another book that may be helpful is *Time Tales: Upper Times Tables in 1 Hour* by Trigger Memory Systems. (Don't believe the part of the title that says in 1 hour, however.) Students are introduced to a cartoon character for each of the numbers. For example, 9 drawn to look like a tree house. Then a short story is given for each of the 6 to 9 times tables. The characters are faded out until just the numbers show.

Some students have learned their multiplication facts from a video or CD made from TV's *Schoolhouse Rock.* The idea came from a father who noticed that although one of his sons was having trouble remembering the multiplication tables, he knew the lyrics to many current rock songs. These songs can be bought from Schoolhouse-Rock.com and other educational sites for about $12 -$15, as well as downloaded from iTunes or borrowed from many libraries.

MAKING YOUR OWN RHYMES AND PICTURES

You may be able to think up or find rhymes on the Internet for specific facts that your student needs to learn and perhaps draw some accompanying cartoon pictures. For example:

- "I ate (8) and I ate (8) 'til I fell on the floor.
  8 × 8 is 64." (See illustration on page 104.)

- Draw 3 trees with a swinging vine between them.
  Swing from tree to tree on a vine
  3 × 3 is the number 9.

- 5, 6, 7, 8
  56 is 7 times 8.

- 6 × 6 = 36 (It rhymes.)

- Hello, hello. How are you?
  6 × 7 is 42."

You will notice that all the strategies for teaching multiplication facts above rely on making the facts more visual or more concrete, or on relating them to easily learned

stories or mnemonics. All of these strategies help hands-on learners, but could make learning the times tables easier for any student.

Games for Multiplication Facts Practice

Games can make practicing the multiplication facts fun. They can be played over and over again, with students using a multiplication table or a calculator at first, until they can automatically state the facts. You might encourage the students to try memorizing the facts by making a new rule that they can use support only every other time or that they can advance one more place on the game board if they can say the fact without help.

TIMES CONCENTRATION GAME

OBJECTIVE: The student will practice the multiplication facts until she is fluent with them.

MATERIALS:
- Cards with the multiplication facts for 2, 3, 4, and 5 from Appendix B
- Cards with the answers to the above facts from Appendix B

PROCEDURE:

1. Divide the multiplication fact cards between the students.
2. Lay down the answer cards in rows (see #6 below) with the faces down. Try to keep the cards in the same places during the game.
3. One student turns over a problem card. She then turns over an answer card. She can use the calculator at first to check what the answer to the problem may be. If she has a matching problem and answer, she puts the two cards together in front of her.
4. If she does not turn over the answer to the card in her hand, she must put the card back, face down, on the table in its former location. Another player then takes a turn.
5. The player with the most matches after 15-20 minutes is the winner.
6. This concentration game also requires the student to remember the location of the answer cards when they have been placed back on the table. It is best to start with only 4 or 6 answer cards at first. Make sure that you deal out the times facts that relate to the specific answers you have chosen. Later you can add more facts.

TIMES COVER

OBJECTIVE: The student will become fluent with the 1-6 times facts.

MATERIALS:
- The Times Cover game board from Appendix B for each player
- Poker chips or tokens (20) that can cover the numbers
- Dice

PROCEDURE:

1. Give each player a game board.
2. Players each roll the dice, and the player with the highest number goes first.
3. The players roll the 2 dice in turn. The player multiplies the two numbers on the dice. (She may use a multiplication table.) The player then covers the answer to her problem on the game board. For example, a player throws a 3 and a 6. She multiplies 3×6 and gets 18. She then covers the 18 on her game board.
4. At first, an adult or older peer should check each answer. Later a player can check the answers with a multiplication table or calculator.
5. The play continues until a player covers any one of the 3 vertical columns. She is the winner.
6. If the players roll the dice 3 consecutive times and are unable to cover any of the answers, the game ends. The players add together the uncovered numbers, and the one with the lowest score wins.

STICK STUFF

Students usually have the most trouble with the facts involving the larger numbers of 6, 7, 8, and 9. They need to practice these fact most. In the game below, the winner is not only the one who provides the most correct facts, but also the one with the sticks having the greatest value. Therefore, winning the game involves chance as well as skill.

OBJECTIVE: The student will practice the multiplication facts from 6 to 9.

MATERIALS:

- Popsicle, craft sticks, or tongue depressors labeled with the numbers 6, 7, 8, 9 (1 set for each player) on one end (use marker or pen)
- 1 soup can (taped around cut end with duct tape) or small plastic cup for each player
- 1 can for the center of the table
- Multiplication table (6-9 facts) for each player (Appendix B)

PROCEDURE:

1. Put the sticks in the center can with the numbers at the bottom.
2. Give each player a soup can.
3. Each player, turning her head away from the can, chooses two sticks.
4. She multiplies the two numbers on the sticks. At first she may use the multiplication table to find the answer. She says the answer out loud.
5. The facilitator (or another peer) checks the answer on the multiplication table.
6. If the answer is correct, the player keeps the sticks and puts them in her can.
7. At a specified time or when the sticks are all gone, the game is over. The players take out their sticks and add all the numbers together. They may use the calculator for the addition.
8. The player with the largest sum is then the winner.

BATTER UP MULTIPLICATION GAME

OBJECTIVE: The student will be able to multiply using the calculator in a game situation.

MATERIALS:

- Cards with 2-3 challenges (math problems) (Appendix B)
- Cards with game calls (Appendix B)
- Baseball game board (Appendix B)
- Markers
- Calculators

PROCEDURE:

1. Put markers on home plate.
2. First player picks a card from the pile (which has been shuffled)

Commercial Games to Buy or Try

- **Mad Math** (made by PATRIX Communications; available from www.mindwareonline.com). Players shake 2 ten-sided dice and multiply numbers 0 × 9. Then they place their marker on the board on the product. (The board is configured so you that can find the product of two numbers by running your finger down from one number and across from the other.) A good game for learning and practicing the times table up to 9 × 9.

- **Flip 4** (available from www.mindwareonline.com). Players roll 2 standard dice and then add, subtract, or multiply the numbers to try to land on numbers on a board.

- **Snap It Up** (card game from Learning Resources). One version is multiplication. Players multiply numbers on the cards they are given, trying to equal "target" cards in the middle of the table.

- **Multiplication War.** Play War with a regular deck of cards (count the ace as 1 and remove face cards). Each player flips over her top two cards, multiplies them together, and says what her product is. The player with the larger product takes all four cards. If players both have the same answer, they each lay down two more cards and the player with the largest product takes all eight cards. (A special **Multiplication War** deck is also available from Barnes & Noble bookstores for about $2.50.)

- **Math Dash** (listed in Chapter 5) can be played using all or only some of the operations (addition, subtraction, multiplication, division).

- **The Winning Touch** (made by Media Materials; available from www. educationallearninggames.com). The game comes with a board that is numbered from 2 to 12 on the top and along the side, and tiles with answers to the multiplication facts from 2 × 2 to 12 × 12. Players begin by randomly selecting 12 answer tiles. They then take turns looking for a place on the board where they can place one of their answer tiles. The catch is that they can only put down a tile if it touches one that is already on the board.  The first player to use up all her tiles is the winner. (Answer tiles for larger fact families such as the 11s and 12s can be removed from game play.)

3. If it is a game call card (one with instructions), she follows the directions, and the play goes to the next person.
4. If she chooses a challenge card (one with problems), she can choose either of the two problems.
5. She solves the problem, with or without using the calculator.
6. She or the teacher finds the answer on the answer key.
7. If she is not correct, she strikes out and leaves the game. Then the play goes to the next player.
8. If she is correct, she may advance one base.
9. Then the other player gets her turn. The object is to keep answering the cards right so you advance around the bases and make a run. Chance, from the game call cards, as well as accuracy in solving the problems, makes a winner. The person with the most runs at the end of play wins the game.
10. The game can also be played with teams, alternating the turns between teams.
11. The game also has a score board that must be marked as the game progresses.

See the Generalization Activities at the end of Chapter 9 for more ideas for assisting students with learning multiplication skills.

Practical Uses of Multiplication

Questions to be answered:

Can the student:

1. Accurately do single- and double-digit multiplication problems using the calculator.

2. Do simple pencil and paper multiplication problems using the times facts the student already knows.

3. Solve simple one-step multiplication word problems by using a structured format
   - Read and understand the problem.
   - Draw or act out the problem.
   - Check out what the operation should be.
   - Write a number sentence.
   - Solve the problem.

Using multiplication in everyday life requires that students be accurate multiplying with the calculator and that they understand when to use multiplication. In this chapter, we will first give the students some multiplication problems that are to be done with the calculator. (If possible, the students will be able to do some of the same problems without the calculator, if they contain facts that they already know.) Then we will discuss how to do some word problems that involve multiplication.

Multiplying with a Calculator

Volume 1 of *Teaching Math* details my rationale for using the calculator early and often with students who are concrete learners. If you have not read Volume 1, refer to Appendix A, for tips on selecting a calculator and on teaching students a procedure to use in entering numbers into the calculator.

INTRODUCING THE CONCEPT

OBJECTIVE: The student will be able to accurately do single- and double-digit multiplication problems using a calculator.

MATERIALS:
- Calculator
- Multiplication by One Digit Worksheet Sheet from Appendix B.

SUCCESS STEP: Ask the student to multiply 2 × 2 with the calculator. Praise correct procedure and answer. If the answer is incorrect, model doing the problem and have the student repeat the problem.

PROCEDURE:
1. Have the student multiply 3 × 6 on the calculator while you watch to see that he is doing it accurately. Do some other single-digit problems until you are confident that the student knows how to do the problems accurately on the calculator. Refer back to Appendix A if the student needs further instruction.
2. Have the student do the first two lines on the Multiplication by One Digit Worksheet.
3. Then have him check those two lines by reversing the order that the numbers are punched into the calculator. The answers should be the same.
4. The student should then finish the practice sheet, checking the answers by reversing the order that he puts the digits into the calculator.
5. Then have the student do 6 problems for calculator practice before you start each following math session.

MULTIPLYING TWO-DIGIT NUMBERS BY SINGLE-DIGIT NUMBERS

If your student has learned the multiplication facts, you will need to teach him how to do problems involving a single-digit times a two-digit number. For this book, any problems larger than that will be done with the calculator.

OBJECTIVE: The student will accurately multiply single- and two-digit numbers, with or without using the calculator.

MATERIALS:
- Whiteboard or slate, with markers or chalk

SUCCESS STEP: Write the number 15 on the whiteboard and ask the student which digit is in the ones place. If he answers incorrectly, tell him the right answer and repeat with a new two-digit number.

PROCEDURE:

1. On the chalkboard or whiteboard, write the problem:

$$\begin{array}{r} 5 \\ \times\ 16 \\ \hline \end{array}$$

2. Remind the student of the place value of the two-digit number. In the problem 5×16, the numeral 1 is in the tens place and the 6 is in the ones (units) place.
3. Explain that in multiplying two-digit numbers, you start with the ones:

$$5 \times 6 = 30$$

4. Next you multiple the tens:

$$5 \times 1 \text{ tens} = 5 \text{ tens or } 50$$

5. Then you add the answers:

$$\begin{array}{r} 3\,0 \\ \times\ 5\,0 \\ \hline 8\,0 \end{array}$$

6. Have the student check the answer with the calculator.

MULTIPLICATION CHECKERS GAME

MATERIALS:
- The checkerboard from Appendix B mounted on a file folder
  OR
- A commercial checkerboard with the math problems written on Post-It notes (trimmed to be smaller than the squares) and placed on the darker squares.
- 24 checkers (12 of one color and 12 checkers of another color)
  OR
- 12 pennies and 12 nickels

SUCCESS STEP: The teacher should lay the checkers (markers) on the dark squares on one side of the checkerboard. Then have the student put the checkers on the other side of the checkerboard. Praise the correct placement of the checkers. If the placement is incorrect, help the student to place the checkers correctly.

PROCEDURE:

1. Use the regular rules of checkers except that in order for a player to move his own piece or jump another player's piece, he must solve the multiplication problem that is on the dark square that he is moving to. Players may use a calculator to solve problems.
2. If a player solves a problem incorrectly, he must stay in place and the other player takes a turn. At first the teacher or other adult should supervise the game and determine whether the answers are correct.

3. If you don't have enough time to finish the game, the player with the most of his opponent's checkers wins.

4. This game can be played over and over again until the players know most of the answers to the problems. Then you can put new problems on the dark squares with Post-It notes, masking tape, etc., or copy the blank checkerboard and write in new problems. The original problems can be easy multiplication facts, the second set of problems can be more difficult multiplication facts, and the third set of problems can be both single- and double-digit problems. If the students enjoy the games, you can make the problems more and more difficult. You will have to make your own answer sheets for the additional problems.

Rules of Checkers

Checkers is a game played between two players who alternate moves. When one player loses all his checkers or all his moves are blocked, the other player wins. A checker piece can move forward one square, diagonally, to a vacant space. A king (see below) can move both forward and backward. A player can capture an opponent's piece by jumping over it, diagonally, to a vacant square beyond it. A player can capture more than one of his opponent's pieces in a move if each of the jumped pieces alternate with vacant squares.

When a piece reaches the last row on the opponent's side of the board, it becomes a king. The opponent places a second (previously captured) checker on top on the piece to "king" or crown it. The turn then passes to the opponent. When one player can't move any checker or when all his checkers have been taken, he loses.

Activities for More Multiplication Practice

Web sites on the Internet can be a good source of problems for the additional practice that a hands-on learner needs to learn multiplication. Some sites provide straight drill of the facts, using electronic flashcards; others incorporate game-like activities. Some sites with colorful, useful activities for students include:

- www.mathforum.org
- www.aaamath.com
- arcytech.org/java
- www.multiplication.com

WORD PROBLEMS USING MULTIPLICATION

Math problems that come up in the course of daily living can often be solved with either repeated addition or with multiplication. However, using repeated addition is not a very efficient way of finding the answer to most problems. I have found that a structured form such as the one used earlier for addition and subtraction problems will help the student know when to use multiplication and to use it more efficiently. The steps are very similar to the addition steps except that the student must notice that the numbers in the problems are the same (equal).

The steps are:
1. Read and understand the problem.
2. Draw, use manipulatives, or act out the problem.
3. Check out what the operation should be.
4. Write a number sentence.
5. Solve the problem.

OBJECTIVE: The student will be able to solve simple one-step multiplication problems by using a structured format.

MATERIALS:
- Sample Word Problems A – D (or A – G), photocopied from text
- Multiplication Word Problem form from Appendix B
- Construction paper to cut into strips or Post-it notes
- Pencil or marker

SUCCESS STEP: Ask the students to cut 2 pieces of paper in half and then cut them into 8 strips each. If the students are unable to cut the strips of paper, you cut them and have the students set them up in piles.

PROCEDURE:
1. Read Sample Problem A below to the students:
   - Four friends are going on a picnic. They each want 3 cookies for dessert. How many cookies should they pack?
2. Put four headings on a whole piece of construction paper:
   Friend 1   Friend 2   Friend 3   Friend 4
3. Label the construction paper strips each—1 cookie
4. Put 3 strips under each friend's heading.
5. Ask the students if the friends need to bring more than 4 cookies or more than 3 cookies (the numbers in the problem). (yes) Emphasize that the answer will be larger than the largest number in the problem.
6. Remind them that to find the answer, they may need to use addition or multiplication, since the answer will be larger than the largest number.
7. Ask if there are equal-sized groups in the problem? (Yes, the friends want the *same* number of cookies.) Emphasize that the fastest way to solve this problem would be multiplication. Do the problem by both addition and multiplication and show that the answer is the same.

8. Give the students the Word Problem form and ask them to fill in the answers from the previous discussion. When you are first introducing the form, write a sentence with blanks for the numbers and a box for the sign of the operation. Show them the completed sample problem A below and discuss the answers.

SAMPLE PROBLEM A
Four friends are going on a picnic. They each want 3 cookies for dessert. How many cookies should they pack?

READ AND UNDERSTAND THE PROBLEM	DRAW (OR ACT OUT) THE PROBLEM
(Write needed numbers with their labels; e.g., 4 desks or underline them in the problem)	Friend 1 Friend 2 Friend 3 Friend 3
(Write answer sentence with a blank)	3 cookies 3 cookies 3 cookies 3 cookies
They need to bring along _____ cookies.	*(Note: The groups above are equal or the same.)*
1	2

FIND OPERATION:	CIRCLE OPERATION
Will the answer be larger than the biggest number in the problem?	
Are the numbers in the problem the same or equal? (Are there equal-sized groups?)	Larger ➤ Addition / equal groups ↓ Multiplication / Smaller ➤ Subtraction / Division
3	4

CIRCLE OPERATION table:

	equal groups ↓
Larger ➤ Addition	Multiplication
Smaller ➤ Subtraction	Division

WRITE NUMBER SENTENCE SOLVE PROBLEM (e.g., 8 + 4 = ___)	WRITE ANSWER SENTENCE *(Write the answer in the blank of the answer sentence in square 1)*
_____ □ _____ = _____	
5	6

SAMPLE PROBLEM B

Alex gives his dog *3 treats* each day. How many treats does he give his dog in a week *(7 days)*?

PROCEDURE:

1. Give the student three strips of paper with the words *dog treat* written on them. Ask the student how many days Alex is going to feed the dog some dog treats. You may underline the numbers in the problem itself or write them in box 1.
2. Have the student fill out each box in the form, assisting him if necessary. Tell the students that when the problem says *each,* it means that each item has the same number of objects in it.

READ AND UNDERSTAND THE PROBLEM	DRAW (OR ACT OUT) THE PROBLEM
(Write needed numbers with their labels; e.g., 4 desks or underline them in the problem) 7 days 3 dog treats each *(Write answer sentence with a blank)* Alex needs to have _____ dog treats for his dog in a week. 1	2

FIND OPERATION:	CIRCLE OPERATION
Will the answer be larger than the biggest number in the problem? Are the numbers in the problem the same or equal? (Are there equal groups of items?) 3	(see table below) 4

CIRCLE OPERATION

	equal groups ⬇
Larger ➜ Addition	Multiplication
Smaller ➜ Subtraction	Division

WRITE NUMBER SENTENCE SOLVE PROBLEM (e.g., 8 + 4 = ___)	WRITE ANSWER SENTENCE *(Write the answer in the blank of the answer sentence in square 1)*
_____ ☐ _____ = _____ 5	6

SAMPLE PROBLEM C

Heidi has 4 flashlights that she uses when the power goes out. They each need 2 batteries. How many batteries does she need in all?

(Remind the students that the word *each* is used to show that each had the same number.)

READ AND UNDERSTAND THE PROBLEM	DRAW (OR ACT OUT) THE PROBLEM
(Write needed numbers with their labels; e.g., 4 desks or underline them in the problem) 4 flashlights need 2 batteries each *(Write answer sentence with a blank)* Heidi needs _____ batteries for her flashlights. 1	 2
FIND OPERATION:	**CIRCLE OPERATION**
Will the answer be larger than the biggest number in the problem? Are the numbers in the problem the same or equal? (Are their equal groups of items?) 3	*(table)* 4
WRITE NUMBER SENTENCE **SOLVE PROBLEM** (e.g., 8 + 4 = ___) ____ ☐ ____ = ____ 5	**WRITE ANSWER SENTENCE** *(Write the answer in the blank of the answer sentence in square 1)* 6

Circle Operation box (square 4):

	equal groups ↓
Larger ➜ Addition	Multiplication
Smaller ➜ Subtraction	Division

SAMPLE PROBLEMS FOR SECONDARY-AGED STUDENTS AND ADULTS

Use the Word Problem Form from Appendix B and give older students or adults problems that may be applicable to their lives. Make several copies of the form.

SAMPLE PROBLEM D

Jon needs 8 quarters each day to ride the bus to and from work. How many quarters does he need for 5 trips?

Jon needs _____ quarters for 5 bus trips.

SAMPLE PROBLEM E

Dara is going on a 5-day vacation to Disney World. She needs to take 4 pills a day for asthma. How many pills does she need to bring?

Dara needs to bring along ____ pills for her trip to Disney World.

SAMPLE PROBLEM F

Kyle and James are making 3 apple pies for Thanksgiving. They need 6 apples for each pie. How many apples do they need in all?

Kyle and James need _____ apples for all the six pies.

SAMPLE PROBLEM G

Julia has a membership at a video store. She is allowed to rent 5 videos a week. How many videos can she rent in one month (4 weeks)?

Julie can rent _____ videos per month.

DETERMINING WHETHER TO USE ADDITION OR MULTIPLICATION

OBJECTIVE: The student will be able to tell when to use addition or multiplication to most efficiently solve a word problem.

MATERIALS:
- Addition and Multiplication Word Problem worksheet from Appendix B
- Slate or whiteboard and chalk or marker
- 9 pencils or craft sticks

SUCCESS STEP: Put 3 pencils, 4 pencils, and 2 pencils in front of the student. Ask if we can multiply to find the answer to how many pencils there are altogether. If the student correctly answers no, rearrange the pencils into 3 piles of 3 pencils each. Then ask if you could multiply to find the answer. Emphasize that the difference is that you can multiply only if the groups (piles) are equal.

PROCEDURE:
1. Do the first problem from the worksheet on the chalkboard as the student works the problem on his paper. Emphasize the two keys that are needed to decide the operation to be used:
    - Is the answer going to be larger than the biggest number in the problem?
    - Are there groups of items that are the same or equal?
2. Go through all the problems on the worksheet. Do not work the problems. Have the students ask the key questions and decide whether the problems would use addition or multiplication. Repeat that one could add all the numbers and get the right answer in the multiplication problems—but it would be much more work. Multiplication is the shortcut for adding equal numbers.
3. Have the students actually work the word problems.

THE JOURNEY GAME

OBJECTIVE: The student will be able to do simple addition and multiplication number and word problems in a game context.

MATERIALS:
- The Journey game—part of original informal assessment from Appendix B
- Die
- Journey problem cards (small numbers) for multiplication and addition—both number and word problems (Appendix B)
- Regular number problem cards for multiplication and addition
- Game markers/pawns

PROCEDURE:
1. Place the game markers on Start.
2. One student throws the die. He moves his marker the number of dots he has thrown with the die.
3. The facilitator hands him the appropriate card (multiplication or addition number and word problems). If the student can solve the problem, with or without the calculator, he moves the marker one place ahead, if the problem is a number problem, and two places, if it is a word problem.
4. The student gets to choose whether he wants a number or word problem.
5. The players must follow the instructions that are given on the spaces where they land.
6. The winner must land on the finish by exactly the right number. He may also win by doing a problem correctly and moving ahead one or two places.

You can control what problems the student gets or you can put the appropriate cards face down on the game board and let the student choose.

The Journey game was very successful with my students! They wanted to play it over and over. I had to make additional addition and multiplication cards so they could play more. It encourages them to try word problems because they get more points for them. However, you can still advance on the game board if you make mistakes and also if you choose all number problems—therefore I could have players at different skill levels. You can also use the game to review addition and subtraction problems.

GENERALIZATION ACTIVITIES

Much of the time your students will be using multiplication in daily life, they will be doubling or tripling an amount. It would be useful for them to be able to mentally double (take 2 x) at least the numbers from 1 -10. If a student can double most of the numbers, he will also be learning the division fact family and that can be very useful for sale prices, etc. (See Chapter 14, Shopping.)

In addition, the following situations might be meaningful for generalizing the use of multiplication in daily life situations:

- Planning and planting flowers or vegetables in a garden (for example, if you want to plant 10 tulip bulbs in 2 rows, how many bulbs do you need?)
- Baking small items such as cookies or muffins (for example, if you bake 2 or 3 cookie sheets with 10 cookies on each, what is the total amount baked?)
- Buying items packed in groups such as drinks or hamburger buns (4 six-packs of bottled water is how many bottles?)
- Figuring fines due for overdue books or videos/DVDs (3 books overdue at 10 cents a day is how much?)
- Figuring out how much allowance money you will receive in a given number of weeks (especially if saving up for an expensive item)
- Computing the amount due for items sold as a fund raising project (You sold 30 chocolate bars at $2 each, so how much money should you turn in?)

Multiplication Terminology in Daily Life

Often when we need to use multiplication in daily life, we are tipped off by special words. For example, an ad may say "Double Coupons," and we are expected to know that we should multiply the face value by two. Try to find some of the following terms in newspapers or magazines and talk about what they mean with your student(s):

- Double—2 times. ("Double cheese" means two times as much cheese. "Double your money back" means two times the amount of money you would usually get back.)
- Triple—3 times
- Twice as much—2 times
- Each—the amount or cost for one item or person

As the opportunity presents itself, also point out usages such as "double play" or "triple play," or see if your student understands what happens in the "Double Jeopardy" round of *Jeopardy* or what you do when you get a double word score in *Scrabble*.

CHAPTER 10

Division

Questions to be answered:

Can the student:

1. Divide a total number of items (under 25) into separate, equal-sized groups.

2. Demonstrate division with concrete and representational materials.

3. Correctly read a division problem and solve it on the calculator.

4. Tell when a remainder is needed in a division problem (rather than show a long string of decimal fractions).

5. Divide with the calculator to two decimal places.

6. Accurately do simple division problems with or without the calculator and check them with multiplication.

7. Recognize the 3 ways of writing division problems.

8. Solve simple division story problems.

In multiplication, equal-sized groups are combined. Since division is the inverse of multiplication, division is separating groups into equal-sized parts. Children often experience division in situations where they have to distribute food, toys, etc. equally among friends or family. For example, a student has a package of five cupcakes to share with 2 friends. She gives herself and her 2 friends 1 cupcake each and finds she has two cupcakes left over. She has divided 3 into 5 and found that the answer is 1 with a remainder of 2.

There are two types of division—sharing division and grouping division. In sharing division, we know the total number of items to be shared and the number of groups and we want to find out how many items per group. For example:

> 10 apples (total number of items)
> 5 friends (number of groups)
> How many apples per person? (2)

In grouping division, we know the total number of items and the size of the groups. We want to know the number of groups that can be served. For example:

> 12 candy bars available
> Each child gets 2 candy bars
> How many children can receive candy bars? (6)

The sharing type of division is used most frequently in daily life. The students need to have many concrete experiences of sharing division so they know when it needs to be used. One way to remember sharing division is by imagining dealing out cards for a game. You know the number of groups (people playing) and the number of cards in the deck. You deal out one card to each player until you have dealt out all the cards with everyone having the same number of cards. Whatever cards are left over is the remainder.

MAT DIVISION

OBJECTIVE: The student will be able to demonstrate division with concrete and representational materials.

MATERIALS:
- Various small items that can be divided into up to 5 groups, such as pennies, buttons, M & M's, straws, pretzels, pencils, etc.
- Five pieces of construction paper to be used as mats (half sheets or full size, depending on the size of the materials)
- 5 brown paper circles to represent cookies, if needed
- Pencil or marker

SUCCESS STEP: Ask the student to put each construction paper mat on the table in front of her working area, counting them as she does so. Have her number the mats from 1-5 with a pencil or marker.

PROCEDURE:
1. Hold up a handful of pennies (or other small objects). Say, "I want to share these pennies between my two friends (not counting me). I am going to put down two paper mats to represent my two friends. How can I make sure each one gets the same number?" The students should be able to say that you can give first one friend a penny and then another friend a penny until the pennies are all on the mats.
2. Have one student distribute the pennies.
3. Give all the students pennies. Tell them that this time they are going to share the pennies between three friends so they should put 3 mats in front of them. To make sure they can divide the pennies evenly, make sure you give them multiples of 3.
4. Monitor whether they can divide the pennies independently.

5. Repeat the sharing division with various numbers of friends and other materials. Tell them that another way of saying "sharing equally" is "dividing equally."
6. Show the following pictures and demonstrate how the sharing can be done on paper.

MAKE IT FAIR

PROBLEM A: You have 12 cookies to share with 4 friends. How many cookies does each person get?

Each friend gets _____ cookies.

7. Give each of the friends a name and number, from 1 – 4 (Julie = 1, etc.).
8. Write a number representing a friend on each cookie and repeat until all the cookies are marked.

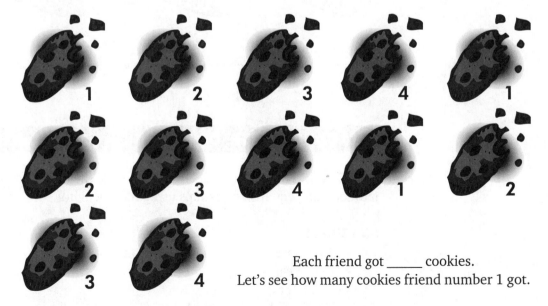

Each friend got _____ cookies.
Let's see how many cookies friend number 1 got.

(Ask them to count how many cookies labeled number 1, then 2, 3, 4 there are.)

3 cookies each

9. If the students have trouble using the picture representation, use brown paper circles for cookies that can be manipulated, or cut out the cookies on Cookie Worksheet (Appendix B).
10. Do the next problem, having the students explain each step as was done in the above problem.

PROBLEM B: There are 4 people in the Jackson family and 8 books to be carried into the library. How many books should each person carry if they share/divide them equally?

Each family member will carry _____ books.

INTRODUCING THE DIVISION SIGN (CALCULATOR ACTIVITY)

OBJECTIVE: The student will be able to divide the total number of items (under 25) into separate, equal-sized groups using the calculator.

MATERIALS:
- 25 or more counting chips (all of one color)
- Deck of cards
- Calculators
- Whiteboard or slate, markers or chalk
- Wrapped candies

SUCCESS STEP: Ask the student to bring you 10+ (an amount that can be shared equally with the number of students in the group) counting chips. Praise her for the correct number or assist her to bring the correct number.

PROCEDURE

1. Using the counting chips that the student has given you, tell the students that you want to divide the chips equally among the people who are there. Only use amounts of chips that can be divided evenly among the number of people there. Deal the chips out as if you were dealing cards.

2. Ask if everyone has the same amount. Tell them that division is about sharing items equally.

3. Give one student some more chips and ask her to share the chips evenly with the number of people there. Make sure that the number can be divided evenly. Then ask how many everyone has received. Say, "Right. We DIVIDED 8 chips into 4 groups and each person got 2…. 8 chips DIVIDED by 4 equals 2 chips each."

4. Teach the students what the division sign looks like: Write the division sign on the slate or whiteboard. Point out the dash with the dot above and below it. Show them the division sign on the calculator and contrast it with the subtraction sign. Also point out the position of the + and the × signs. If your student has trouble finding the division sign, put a small dot of tape on it on the calculator.

5. Repeat the above problems using the calculator. Say, "I have a total number of ___ chips. I always enter the total number in the calculator first. Then I divide (push in the division sign on the calculator) that number by the number of people here. Then I push the equals sign. Is my answer the same as what we got when we distributed the chips to everyone?" For example, 10 (total number of chips) divided by 5 (number of people) = 2 chips each (modify the numbers for the number of chips and people in the room).

6. Pair off the students, if possible. Give one of the partners some more chips.

7. Have that partner divide the chips between herself and her partner. The first person should check the division by putting the problem in the calculator and solving it. At this time put any leftover chips (remainders) down on the floor.

8. Teach the students the chant for entering division problems in the calculator and have them practice putting in problems in the calculator as they chant.

> The biggest number goes in first.
> *(Spread your arms out wide as if holding something big.)*
> Then the division sign.
> Put the number sharing.
> And press the equals sign.
> We have our answer met.
> How many do **I** get? *(point to self several times)*

You can sing the first 4 lines to *Row, Row Your Boat,* then just *say* the last two lines with emphasis. You will need to repeat the chant (song) at every session until the words become automatic.

9. Now, give each pair of students an even number of wrapped candies. Have one of the pair divide the candies and the other person check the problem on the calculator

with both students doing the chant. If you do not want to use candy, use individual stickers, small erasers, marbles, or sticks of gum as an alternative to candy.

10. Adjust the number of wrapped candies that the pair holds and have the partners reverse jobs.

11. Repeat as often as is necessary.

12. Let each of the students sample one of the candies—and save the rest for a snack.

13. At another session see if the students can come up with experiences that require sharing division such as sharing a batch of cookies or M & M's.

INTRODUCING THE DIVISION HOUSE (CALCULATOR ACTIVITY)

The language describing division can be difficult for students with Down syndrome or other language delays:

$$6 \div 3 \text{ is read } \textbf{6 divided BY 3.}$$

$$3\overline{)6} \quad \text{is often read } \textbf{3 into 6} \text{ or } \textbf{Divide 6 BY 3}$$

OBJECTIVE: The student will be able to correctly read a division problem written two ways and solve it on her calculator.

MATERIALS:
- A calculator for each student
- Simple Calculator Division Problems from Appendix B
- A slate or whiteboard and chalk or marker
- Sheets of paper and a pencil for each student and teacher

SUCCESS STEP: Write $8 \div 4 = 2$ on the slate or chalkboard. Ask the student to read it aloud. Repeat her words if she is right. "Yes, this says 8 divided by 4 equals 2." If not, have her model your words.

PROCEDURE

1. Write the following problems on the whiteboard:

$$6 \div 3 \qquad 3\overline{)6}$$

2. Tell the student that these are two different ways of writing the same problem. In the United States the $\overline{)}$ is often used to show the process of division. We will call this a "division house." The number to be divided (dividend) is placed inside the division house. The number to be divided by (divisor) is placed outside the house at the door. The answer (quotient) is placed on the roof of the house. Draw the division house on the slate or whiteboard and explain the parts.

3. Write a large blank division equation on a piece of paper using a division house.

$$\underline{\quad}\overline{)\quad\quad}$$

4. Show 12 poker chips on the table. Ask the students how you can divide these chips among three people. After they respond, show them another way: Draw 3 circles (labeled 1, 2, and 3) on the paper inside the "division house." Have a student put one chip in each circle, then distribute all the chips in the circles. See how many chips are in each circle. Write the number on top of the division house. Write the problem 12 ÷ 3 = 4.

5. Now model how you can do the same problem with dots on the paper instead of poker chips. You can model putting 12 dots on the paper and then drawing the "division house" over the group of 12 dots. Put the number you are dividing by (3) outside the "division house." See how many groups of 3 each you can make out of the 12 dots and circle each group. Count the number of groups that you have made and put that number on top of the "division house."

6. Now do the same problem on the calculator. Remind them that they have to put the largest number in the calculator **first.** Say the chant:

> *The big number goes in first*
> *Then the division sign*
> *Put the number sharing*
> *And press the equals sign.*
> *We have our answer met.*
> *How many do **I** get?*

7. Have the students do the page of Simple Calculator Division Problems from Appendix B. All of the problems will come out even. Remind the student that if the problem is written in the division house form, the big number will be inside the house.

Remainders

For division, as for the other math operations, *Teaching Math* advocates teaching students to rely primarily on their calculators, rather than mental math. Using a calculator for division, however, requires a judgment call that is not needed when doing the other operations. That is, when the answer does not come out even, you

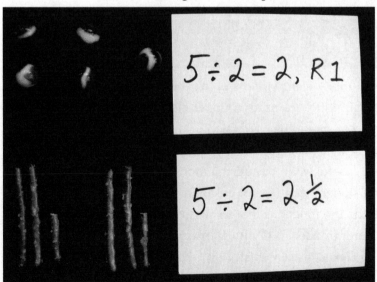

have to decide whether to express the answer as a decimal or a remainder. We have not introduced decimals yet, except where money is involved, so decimals will be meaningless to the students at this point.

For now, when you solve a division problem that results in a remainder, explain to the students that the calculator doesn't know how to do remainders. It just goes on and on with numbers that are smaller than one whole number. For example:

> *Mrs. Meyer bought a bag of balloons for her 4 children to share. There are 10 balloons. How many can each of her children have? (2 balloons)*

Can you divide the leftover two balloons equally between 4 children? (Not if we want to put air in them.) However, if you do the problem on the calculator, the answer comes up 2.5. For now, we will drop the .5 and say the answer is 2 with a remainder.

SHOW YOUR REMAINDERS

OBJECTIVE: The student will be able to tell when she needs to show a remainder.

MATERIALS:
- 10+ pencils
- Small candies or other small items that cannot easily be divided such as M & M's, Reese's Pieces, marbles, or erasers (enough to give 3 to every person, plus one extra)
- Items that can be divided such as licorice sticks or straight pretzels

SUCCESS STEP: Ask one student to divide an amount of pencils evenly among the other students or the rest of the family. Say, "Divide up the pencils evenly. Give me back any pencils that are left over." Praise her for doing it correctly. If she is incorrect, model the actions for her, collect the pencils and have her try again.

PROCEDURE:
1. Show the students wrapped candies (1 more than 3 candies for each person). Tell the student(s) that you want to be fair and give everyone the same number of candies.
2. Deal out 3 candies per person with one left over. See if they can tell you what the remainder is going to be before you deal out the last set of candies.
3. Ask the students what to do with the one left over. Can you cut it easily into pieces? Hopefully, they will answer no.
4. Explain to them that you just have one candy left over. It is a remainder.
5. Let them eat their candies or give them the marbles and tell them that you are saving the remainder for an absent classmate, sibling, or parent.

Time for a New Calculator?

If your student is able to use a calculator with many extra buttons on it, you may want to buy her a fraction calculator. These usually cost about $20 and are used in upper elementary and secondary schools. These calculators can be programmed to give either a fraction or a remainder as the answer to division problems. Some also have displays that can show the whole problem as it is being solved or let you set the number decimal places to show (useful when calculating money amounts and you want to show two places). However, if you give the student that type of calculator, you must figure out the instructions for yourself and teach her how to do the functions that will be useful for her!

Some useful models are the Math Explorer (model TI-12, Texas Instruments), the Fraction (Casio), and the Explorer Calculator (model TI-15, Texas Instruments).

6. Repeat the procedure with items such as pretzels that can be divided. Distribute the pretzels so each person has 2 but there are enough left over to give each person ½ of a pretzel. See if you can get the students to come up with the idea that the pretzels could be broken in half and everyone else could have an equal amount. Later, distribute a different (uneven) number of pretzels to see what the students will do with the one that is left over.

7. Make the point that sometimes you just need to leave a remainder and sometimes you can further divide the items so each gets an equal part.

DIVISION TIC-TAC-TOE GAME

OBJECTIVE: The student will correctly solve simple division problems in a game situation.

MATERIALS:
- Division Tic-Tac-Toe game from Appendix B
- Magnetic bingo chips (two different colors) or other markers (you could use pennies and dimes)
- Magnetic bingo wand (from discount store or Internet)
- Calculators

PROCEDURE:
1. The rules are similar to tic-tac-toe except that the students need to solve the division problem on the square that they choose before they can place their marker on it. They may use a calculator.
2. Each person puts her marker or token on a square. She must solve the division problem on the square to be able to stay there.
3. The first person to put her marker on three squares in a row or diagonally wins that game. That person gets to use the magnetic Bingo wand to pick up her Bingo chips first. The other players get to pick up their Bingo chips when she is done.
4. You can make your own games by writing your own problems on the blank form.

Long Division

As early as 1981, a textbook on teaching mathematics suggested the feeling of many mathematics educators:

> *"Increased availability and lower cost of hand calculators, together with the infrequent need to use long division in day-to-day situations, has caused mathematics educators to question whether we should continue to teach the long-division algorithm" (Bley & Thornton, 1981).*

Several years later, the National Council of Teachers of Mathematics (as quoted in Reys, Suydam & Lindquist, 1984) noted:

> *"For most students, much of a full year of instruction in mathematics is spent on the division of whole numbers—a massive investment with increasingly limited productive return.... For most complex problems, using the calculator for rapid and accurate computation makes a far greater contribution to functional competence in daily life."*

To do long division, we have to use several different processes and concepts, for example:

$$56 \div 4$$

1. **Estimate** how many times the divisor (4) can be divided into the dividend (56): 5 (tens) divided by 4 = 1 (tens)
2. **Multiply the** 1 times the 4
3. **Compare** remainder
4. **Subtract** the 4 from 5 to get 1 (ten)
5. **Bring down** the 6 (ones)
6. **Divide** the 4 into 16 = 4
7. Then **multiply** 4 × 4 = 16
8. **Compare**
9. **Subtract**

Most students with Down syndrome and other hands-on learners have great difficulty estimating, remembering so many steps, lining up the numbers, etc. Therefore, in this book we will only discuss using a calculator for long division. Showing remainders as decimal fractions will be addressed, however, because of their use with money (see Chapter 13).

USING A CALCULATOR FOR LONG (MULTI-DIGIT DIVISORS) DIVISION

OBJECTIVE: The student will be able do multi-digit division with the calculator.

MATERIALS:
- Calculators
- Calculator Division Problems (Appendix B)

SUCCESS STEP: Ask a student to do a simple one-digit division problem, using the calculator. Praise her successful answer or help her to correct the problem.

PROCEDURE:
1. Go through the first problem on the worksheet with the students.

2. Have them do the second problem and check it before assigning the sheet to be done independently.
3. Have the students complete the Calculator Division Problems from Appendix B.

What about Division Facts?

If the student has learned most of the multiplication facts, she can probably learn the associated division facts. For example, if the student knows the 4s multiplication facts, she can probably learn to reverse them to come up with the division facts. Think of how you use the basic math facts. Many adults recall the multiplication fact when doing division. For example, in the problem 42 ÷ 6, most adults, think "what number times 6 equals 42?" Then they recall the multiplication fact of 7 × 6 = 42 rather than recalling a division fact.

REVERSING MULTIPLICATION PROBLEMS

OBJECTIVE: The student will be able to demonstrate with objects that division is the reverse of multiplication.

MATERIALS:
 • Bingo chips or poker chips (20 per student)

SUCCESS STEP: Ask the students to answer a simple multiplication problem (like 3 × 4 = 12). Keep giving examples until you have a easily recalled fact (answer under 20).

PROCEDURE:
1. Start with the multiplication fact that the students recalled in the Success Step.
2. Have them show you how to represent that fact with bingo or poker chips (making 3 groups of 4).
3. Ask them to tell you how many total chips there are (12).
4. Then say, "What if we DIVIDED these same 12 chips into 3 groups? How many chips would we have in each group?" See if they realize that the answer is already right in front of them.
5. If not, talk them through it… "How many chips total do we have here? (12) How many groups of chips are there? (3). So, if we divide 12 chips into 3 groups, how many chips are in each group?"
6. Show them on a whiteboard how multiplication is the reverse of division:

$$3 \times 4 = 12 \qquad 12 \div 3 = 4$$

7. Then do more examples of multiplication sums that they know very well, modeling with chips and seeing if they can understand that you should know the answer to the division problem if you can do the multiplication problem.

DIVISION WORKSHEETS

OBJECTIVE: The student will accurately do simple division problems with or without the calculator and check them with multiplication.

MATERIALS:
- Division Problems Checked by Multiplication worksheet (Appendix B)
- Pencil
- Calculator

SUCCESS STEP: Ask the student what kind of problems are shown on the worksheet. Praise the answer.

PROCEDURE:
1. Have the students do the first problem on the worksheet. Then compare answers.
2. Show the students how to check division problems. That is, after the students have solved the problem, have them multiply the two smaller numbers (the divisor and the quotient) to find the big number (dividend). For example,

$$16 \div 4 = 4; 4 \times 4 = 16$$

3. Have the students do the entire Division Worksheet and check each problem with multiplication.
4. Give the students more practice in division using the other division worksheets.
5. Many mass-market workbooks have simple division problems that can give more practice to the students. You can also make simple division problems yourself. At first, write simple problems that have no remainders so they can be easily checked by multiplication.

Division with Story Problems

Students need to have real-life and simulated experiences with sharing and grouping division. Some experiences will be simulated now and repeated when all types of story problems are discussed.

SAMPLE PROBLEMS A, B, AND C

OBJECTIVE: The student will be able to solve simple division story problems.

MATERIALS:
- A bag with 20 marshmallows in it
- Word Problem form (4 forms for each student) from Appendix B.
- Post-It notes, if needed

- 35 straws or craft sticks per student
- Calculators

SUCCESS STEP: Have the student count the marshmallows that are in the bag. out loud. Praise correct counting.

PROCEDURE:

1. Read Sample Problem A with the students.

SAMPLE PROBLEM A:

Emily's family wanted to roast marshmallows. There were 20 marshmallows left in the bag. There were 4 people in Emily's family. How many marshmallows could each person get? (if they were divided evenly)

2. Have the students work out the first problem, step-by-step, with you.

   a. **Read and understand the problem** (See Chapter 6, addition and subtraction word problems if you have not used the form before.)
      - Write the numbers of the problem—
        4 in family, 20 marshmallows

      - Write the answer sentence (subject, verb, object) with a blank for the answer.
      Each family member could have ____ marshmallows to roast.

   b. **Draw (or act out) the problem** (See step e, below.)

   c. **Find the operation:**
      - Will the answer be **larger than the largest** of the numbers in the problem?
      No, the largest number is 20 and no one in Emily's family could have more than the total number of marshmallows left in the bag.
      Therefore, the operation will be either subtraction or division.

      - Are the numbers for each person in the problem the **same or equal**? That is, do we need to make equal groups?
      Yes, each person will get the same amount. Therefore, the operation is division.

   d. **Solve the problem.**
      Using the calculator, enter the 20 marshmallows divided by 4 (number in family) = 5 marshmallows each. Remember to put the largest number in first.

   e. **Check the above problem** by choosing 4 people and handling out real marshmallows or construction paper circles to represent marshmallows. If you do not have 4 people, use Post-It notes to represent each member of the family. Deal out the marshmallows to each family member and show that each member gets 5 marshmallows.

3. Use the word problem form that has been used with multiplication in Chapter 7. Go through each step in sample problem B as was done in sample problem A.

SAMPLE PROBLEM B:
LaNell bought 18 American history stamps on her trip. If she gives the *same* number of stamps to each of her 6 friends, how many stamps will each friend get?

READ AND UNDERSTAND THE PROBLEM *(Write needed numbers with their labels; e.g., 4 desks or underline them in the problem)* (Write answer sentence with a blank) 1	**DRAW (OR ACT OUT) THE PROBLEM** 2

FIND OPERATION:

Will the answer be larger than the biggest number in the problem?

Are the numbers in the problem the same or equal? (Are their equal groups of items?)

3

CIRCLE OPERATION

	equal groups ⬇
Larger ➜ Addition	Multiplication
Smaller ➜ Subtraction	Division

4

WRITE NUMBER SENTENCE
SOLVE PROBLEM (e.g., 8 + 4 = ___)

5

WRITE ANSWER SENTENCE
(Write the answer in the blank of the answer sentence in square 1)

6

4. Try sample word problem C using the structured form provided.
5. Have the students use manipulatives such as straws to check sample problems B and C.

SAMPLE WORD PROBLEM C:

Jake took 33 photos on his trip. If he divides them *equally* between 3 friends, how many photos does each friend get?

READ AND UNDERSTAND THE PROBLEM	**DRAW (OR ACT OUT) THE PROBLEM**
(Write needed numbers with their labels; e.g., 4 desks or underline them in the problem)	
(Write answer sentence with a blank)	
1	2

FIND OPERATION:	**CIRCLE OPERATION**

FIND OPERATION:

Will the answer be larger than the biggest number in the problem?

Are the numbers in the problem the same or equal? (Are their equal groups of items?)

3

CIRCLE OPERATION

	equal groups ↓
Larger ➜ Addition	Multiplication
Smaller ➜ Subtraction	Division

4

WRITE NUMBER SENTENCE **SOLVE PROBLEM** (e.g., 8 ÷ 4 = ___)	**WRITE ANSWER SENTENCE** *(Write the answer in the blank of the answer sentence in square 1)*
5	6

6. Photocopy the following problems from A through F.
7. Have the students do the problems independently using the preceding structured form or copy the word problem form in Appendix B:

    a. Paul read a 60-page book on his vacation. He read 10 pages *each* day. How many days did it take him to read the whole book?

    b. Caroline has baked 28 chocolate chip cookies. She is going to give them to 8 Cub Scouts to eat. How many can each Scout get (if she gives them each the *same* amount)?

    c. Heidi bought 24 potted flowers. She and her two children planted the flowers in the front yard. If the 3 of them each planted the same number of flowers, how many flowers did each person plant?

    d. Declan ordered a pizza that was cut into 12 pieces. Counting Declan, there are 4 guys who want some pizza. How many pieces of pizza should each guy get—assuming they each get the same amount?

    e. Katri bought some large beads to make a bracelet. The package says that there are 40 beads in it. She thinks that she may have enough beads to make 2 bracelets. If she separates the beads into 2 equal groups, how many beads will she have for each bracelet?

    f. Jeff has 15 tools. He wants to put an equal number of tools in each of his 3 toolboxes. How many tools should he put in each toolbox?

THE JOURNEY GAME

OBJECTIVE: The student will be able to do simple subtraction and division number and word problems in a game context.

MATERIAL:
- The Journey game—part of original informal assessment from Appendix A
- Die
- Problem cards for subtraction and division – both number and word problems (Appendix B)
- Game markers

PROCEDURE:
1. Place the game markers on Start.
2. One student throws the die. She moves her marker the number of dots she has thrown with the die.
3. The facilitator hands the appropriate card (subtraction or division number and word problems) to the student. If the student can solve the problem, with or without the calculator, she moves the marker one place ahead, if the problem is a number problem, and two places, if it is a word problem.
4. The players must follow the instructions that are given on the spaces where they land. If a player lands on an instruction space a second time, she does not have to obey the instructions.
5. The winner must land on the finish by exactly the right number. She may also win by doing a problem correctly and moving ahead one or two places.

6. You can control what problems the student gets or you can put the appropriate cards face down on the game board and let the student choose.

GENERALIZATION ACTIVITIES:

- One of the most common uses for division in everyday life is distributing a group of items equally (sharing). Teachers and families can label these situations as *making it fair* by division. Examples include:
  - ○ Handing out food such as cookies or candy
  - ○ Dealing out cards in various games
  - ○ Passing out arts and crafts or school supplies
  - ○ Dividing up tickets for carnival rides or arcades
- When your child asks you how many slices of pizza, pieces of bread, muffins, ears of corn, etc. she can have at a meal, have her do the division: "There are 8 slices of pizza and 4 of us…. If we share the pizza fairly how many should we each get?"
- When shopping, involve your child in division scenarios. "There are 8 hamburger buns in this package…. Is that enough for us? (Yes, if you are dividing among 4 family members; no, if you are having a cookout with 12 guests.)
- Divide up minutes of computer, video game, TV time among family members. "There are 40 minutes before we have to leave and you both want to use the computer. 40 minutes divided by 2 means you can each use the computer for 20 minutes."
- At school or home, divide tasks by number of days. For example: "We need to learn 15 new words this week. How many do we have to learn a day?" Or, "You need to read 20 pages in 4 days. How many pages should you read a day?"
- At school, when dividing students into groups, figure out how many people there should be per group: "We need to make 3 groups. There are 20 students. How many per group?"
- If the student enjoys downloading songs from the Internet using an iTunes or other gift card to pay for the downloads, help her understand how to use division to figure out how many tunes she can download per card. For example, if each tune costs $.99 to download, explain that that is basically $1, using the next-highest-dollar strategy. With a $15 gift card, you divide by 1 to see how many songs you can buy.

Commercial Games to Buy or Try

- **Division War:** Barnes and Nobles bookstores sells decks of cards with division facts for playing the popular game War. All the cards are dealt out to players. Then players flip over the top card in their stack and announce the answer of their division problem. The player with the highest answer takes the other players' cards. (You can also make your own division war deck with index cards.)

- **Multiplication & Division Quizmo** (World Class Learning Materials): A bingo-like game where one player reads out division (or multiplication) problems and the other players check their boards to see if they have the answer. If they do, they cover it with a plastic chip. The first player to fill in a line horizontally, vertically, or diagonally is the winner.

CHAPTER 11

Simple Fractions

Questions to be answered:

Can the student:

1. Recognize one-half of an object and read the fraction as one-half (½) when written.

2. Read the fraction ¼ and demonstrate its meaning.

3. Use halves and fourths in baking.

4. Read the fractions ⅓ and ⅔ and demonstrate their meanings.

5. Use thirds in baking.

6. Read the common fractions (½, ⅓, ¼, ⅕, ⅙, ⅔, ¾, ⅖, ⅗, ⅘, and ⅚) correctly.

7. Show that ½, 2/4 (and 4/8) are the same amount.

8. Demonstrate that when a fraction has the same number on the top and the bottom, it means that the fraction is really the whole number 1.

9. Read mixed numbers (1- to 2-digit whole numbers and simple fractions) as well as demonstrating understanding with real objects.

10. Add fourths together and later add thirds.

11. Use fraction strips to demonstrate and compare common fractions.

12. Bake—using whole number fractions and mixed numbers.

Children often come to school knowing generally what half of an object is. They usually seem to realize that both halves need to be the same size—or at least they will protest if their half of something desirable is smaller than the part given to another child. We can work from that simple concept to broaden their understanding of fractions that are necessary in daily life.

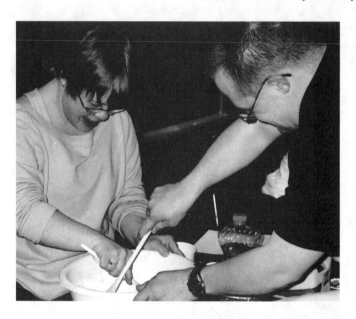

First we will work with common or simple fractions expressed as one number on top of another number such as $\frac{1}{2}$ (or, as typed, ½ or 1/2). Initially, the students need to be able to read common fractions accurately. Then they need to be able to read mixed numbers that contain a whole number and a fraction, such as 2½.

How are fractions commonly used in everyday life? Fractions are most often used in measurement. Cooking, using fractions of teaspoons, tablespoons, and cups, is probably the most common practical application of simple fractions. Measuring length or weight may also require the accurate use of fractions. Even if your student is included in a regular math class and will be expected to learn more complex math involving fractions, I recommend that you introduce him to fractions through the hands-on activities in this chapter. If a student has a concrete, visual understanding of what a fraction actually means, he is more likely to understand why the denominator stays the same when you add ¼ plus ¼, or why it makes no sense to add ½ to ¼ and get ⅙.

Different Styles of Writing Fractions

Fractions are written differently depending on the printing or computer word processing style. We usually handwrite a fraction by putting one number on top of a line and the other number below the line; e.g., $\frac{1}{2}$. Handwritten fractions usually have smaller-sized numerals than whole numbers do because the two numerals in the fraction are written in the same space as one regular numeral is written. The same is sometimes also true of printed fractions. This difference in size can help students understand that fractions represent smaller quantities than regular whole numbers do.

In some computer and printing fonts, the first number (numerator) is printed just like a whole number, followed by a slash and then the denominator; e.g., 1/2. Some word processing programs automatically change the 1/2 to ½. Therefore, fractions can be written different ways in the same text. Your students need to learn that the different ways of printing do not change the value of the fractions. Be sure to show them visually how the fractions can be written differently, but named and used the same—depending on how they are written in the student's text or tests.

Introducing Halves and Fourths

HALVES

OBJECTIVE: The student will be able to recognize one-half of an object and read the fraction as one-half (½) when written with numerals.

MATERIALS:
- Apples, pears, or some other fruit that can be divided into halves
- Other rewarding whole objects such as candy or granola bars
- Sharp knife
- Construction paper
- Slate, whiteboard, or tablet of paper
- Marker or chalk

SUCCESS STEP: Cut the fruit in half and put both halves in front of one student. Ask the student to give you one-half. Praise him for correct action or help him give you one-half of the fruit.

PROCEDURE:
1. Show your student(s) the half fruit used in the Success Step and ask him how much of the fruit this is (one-half). Then take another fruit and cut it so one of the pieces is much smaller than the other. Hold up the smaller piece and ask the student if this is one-half of the fruit. He should reply that the smaller piece is not one-half because the pieces are not the same. If he does not reply correctly, compare the smaller piece with half of the original fruit that you cut previously.
2. Write the fraction $\frac{1}{2}$ on a slate or a whiteboard. Point out that the fraction is written with a 1 on the top and the 2 on the bottom but we call it *one-half*. Have students repeat the words *one-half* several times while looking at the written fraction. Show them on the board that sometimes one-half is written 1/2 because it is hard to type ½. They should practice reading $\frac{1}{2}$, ½, and 1/2 as one-half.
3. Tell the students that you are going to give one-half of a fruit (or candy bar or any whole object) to each person and they are to watch you so you give them each one-half. Cut fruit, granola bars, or candy bars, etc. in half and give them the halves. Make a "mistake" and cut one of the objects very unevenly. The students should protest that cut if they understand what one-half is. Cut another object correctly to give them equal halves. If you have only one student, you will have to have several different objects to cut in half.
4. Review that $\frac{1}{2}$, ½, and 1/2, is read as one-half.
5. Give each student a piece of construction paper and scissors and ask him to give you one-half of his paper. If necessary, you may cut a piece of construction paper in half to give them a model.
6. Use other things in their environment that can be divided in half (pieces of chalk, pretzel rods, sandwiches, pieces of licorice, fruit leather, lengths of ribbon or string) and give the students chances to make two halves and give you one-half. Remind the students that breaking an object in half is the same as breaking it in the exact middle.

When Kayli was reading a list of common fractions, she correctly read ⅙ as one-sixth and ⅐ as one-seventh, but she read ½ as "one twoth" and ⅓ as "one threeth." She had correctly figured out the pattern for naming fractions from ¼ to ⅑ and applied the pattern to ½ and ⅓. She needed to be taught how to read ½ and ⅓ as separate fractions that do not fit the pattern.

FOURTHS

OBJECTIVE: The student will be able to read the fraction ¼ and demonstrate its meaning.

MATERIALS:
- Three pieces of construction paper for each student
- A marker or crayon for each student and the teacher
- A slate or a tablet of paper for the teacher
- Approximately 1 cup of uncooked rice for each student placed in a large bowl on the table
- An empty bowl for each pair
- A one-cup measuring cup
- A ¼-cup measuring cup

SUCCESS STEP: Put the construction paper on the table in front of the student and ask him to choose *one* piece of paper. Praise the student for only taking *one* piece of paper.

PROCEDURE:
1. Explain that each student has *one* big sheet of paper. Say, "I am going to make my *one* piece into smaller pieces."
2. Fold and tear the paper into fourths. Say, "I have 4 pieces of paper now. Is each piece of small paper as big as your ONE BIG paper?" (no)
3. Put the 4 pieces down on the table and fit them together into ONE BIG piece. "I have 4 parts of this ONE BIG piece of paper—1, 2, 3, 4." Pick up one of the smaller pieces and say, "Now I am holding 1 piece of the 4 pieces of paper. This piece is called *one-fourth*."
4. Write the fraction ¼ on the slate and point to the 1, telling them that it stands for 1 piece out of the 4 pieces that make up the BIG piece of paper. Point out that the numeral 4 stands for the 4 pieces that make up the parts of the BIG piece of paper.
5. Ask the students to fold and tear their paper into fourths as you did.

   Note: A student could also use scissors or someone else could tear the creased lines if he is unable to tear it evenly.

6. Ask students, in turn, to give you one-fourth of their piece of paper. Say, "Thank you for the **one-fourth** piece of paper."
7. Then give each piece back to the student, repeating "I give you back your **one-fourth** piece of paper." Use enough repetition so the students can repeat the term *one-fourth* easily. Write the fraction on the slate several times.

8. Review the fraction ½ by tearing a piece of paper in half and contrasting it with the ¼ pieces of paper that they have made. Tell them than the bottom number of a fraction tells you how many pieces or parts there are in all.

9. Show the students a one-cup measure and a ¼-cup measure. Using the rice, show them that four ¼ cups of rice fit in the one-cup measure. Point out that the ¼ cup only holds one-fourth as much as the whole cup.

10. Have the students put four ¼ cups of rice into the one-cup measure.

MORE FOURTHS

OBJECTIVE: The student will be able to read the fractions ²⁄₄ and ³⁄₄ and demonstrate their meanings.

MATERIALS:
- A piece of construction paper for each student and the teacher
- A slate or a tablet of paper for the teacher and chalk or marker
- 1 teaspoon and ¼ teaspoon for each pair of students
- 2 or 3 cups of uncooked rice in a bowl to be used for the group of students
- An empty bowl for each pair
- Three different colored full sheets of paper (for each student) labeled ¼, ²⁄₄, and ³⁄₄ on top of each piece
- A one-cup measure and a ¼-cup measuring cup for each pair
- Three clear glasses or bottles for each pair

SUCCESS STEP: Ask the student to fold and tear a piece of construction paper into fourths as he has done previously. Praise his success or assist him, if necessary.

PROCEDURE:
1. Have all students fold and tear the construction paper in fourths. Ask them to give you *two-fourths* of their colored sheet of paper. Then repeat with them giving you *three-fourths* of the paper.

   Note: Do not attempt to teach them that ²⁄₄ equal ½ at this time. If a student who already knows about ½ figures it out, praise him, but do not explain at this time.

2. Pair the students up and have them ask each other for ¼, ²⁄₄, and ³⁄₄. If one student is not able to speak clearly, write the fractions on the slate and have him point to the fraction that he wants the other student to give him. After each pair is finished, let each one gently let the pieces of paper fall on his partner's head, as a joke-type reward.

3. Give each pair of students one teaspoon. Remark that this teaspoon is sort of like the spoon you use for eating cereal.

4. Then give the students a ¼ teaspoon. Tell them that this ¼ teaspoon is *really* a small spoon. Ask them if they can guess how many of the small ¼ teaspoons can fit in the 1 teaspoon. Have one member of the pair scoop up ¼ teaspoon of rice and put it in the 1 teaspoon. See if the students can put four ¼ teaspoons into

the 1 teaspoon. They will have to have *level* one-fourth teaspoons to fill it exactly. Empty the teaspoon.

5. Have the other member of the pair scoop up ²⁄₄ teaspoons of rice and put it in the bowl. He should fill the ¼ teaspoon two times. Empty the teaspoons. Then the first person should put in ¾ teaspoon by filling the spoon 3 times and putting the rice in the bowl. Reverse the partners and repeat.

6. Give each student three pieces of white paper which are labeled ¼, ²⁄₄, and ¾. Direct each student to put ¼ teaspoon of rice on the piece of paper labeled ¼, ²⁄₄ teaspoon on the paper labeled ²⁄₄, and ¾ teaspoon on the paper labeled ¾. The teacher should be able to check these answers visually.

7. Ask the students which is the largest pile of rice. Point out that ¾ teaspoon of rice is larger than ¼ teaspoon of rice. Write the fractions ¼, ²⁄₄, and ¾ on the board or pad of paper. Show them that if several fractions have the same bottom number, you can tell which fraction is larger by looking at the top number of the fraction. Put all the rice back in the central bowl.

8. Have the students repeat the procedure in steps 4 and 5 using a ¼ *cup* measure and three clear glasses or bottles. Use a funnel, if it helps. Ask the student which glass has ¼, ²⁄₄, or ¾ of a cup of rice. He should be able to tell you by looking at the glasses.

GENERALIZATION ACTIVITY

One way that you can add interest to visualizing fractions is to use the "Hershey Method." A chocolate Hershey bar has indentions dividing it into twelve parts (4 columns of 3 pieces). The parts are rectangular and look like the fractions the student has been making by tearing or cutting up pieces of construction paper. The parts can easily be broken to show fractions of the whole bar. The teacher needs to lead the group (or individual) in discovering ½'s, ⅓'s, and ¼'s. It is also possible to see equivalents such as ½'s and ⁶⁄₁₂'s. (There is a book by Jerry Pallotta called *The Hershey's Milk Chocolate Bar Fractions Book* that can be used to introduce the concept. It's currently out of print, but you can find copies in online and used bookstores and in the library.) Of course, the students will enjoy the method more if they get to eat the results when the lesson is over.

Baking to Teach Fractions

It is important that students who are hands-on learners learn about fractions in a meaningful way, and baking is a good way for them to learn. You will have to adapt the use of the recipes to your home or school situation. I have included some simple recipes using the first few fractions that are introduced. You do not have to use these recipes. You can add your own recipes or those that the students really enjoy. If you do not have an oven available, you can use one of the microwave recipes or one that uses a waffle iron. If you are unable to cook in any way in your school, you can try the recipes for smoothies in Appendix C.

I have tried to avoid some of the more difficult cooking terms such as "cream" and have not used the abbreviations for teaspoon and tablespoon. Your concentration should be on the fractions, not the cooking. Later, you may want to do other recipes using cooking terms and the standard cooking abbreviations. You will have to explain these terms if the students are to independently use standard recipes.

At first have a student do each step with close supervision. If you are working with a small group, use a different student for each step and have the rest watch. Later on you may be able to assign tasks and have them work without much supervision.

BAKING WITH FRACTIONS

There are some ingredients that are used in many recipes. The following are supplies that I keep available:

- White flour
- White sugar
- Brown sugar
- Eggs
- Milk or water
- Cooking oil
- Soft margarine
- Vanilla
- Baking soda and powder
- Cooking spray such as Pam

I put the flour and the white and brown sugar in refrigerator storage containers that have a sealable top. The students seem to handle the ingredients better if they are not in large quantities such as 5 lb. of flour. Before each cooking session, I buy the special ingredients such as nuts, chocolate chips, blueberries, etc. that are needed for a particular recipe.

The students need to learn how to pour or spoon the ingredients into the measuring utensils. They also need to know how to make the ingredients level in the cup measures such as by leveling off the heaping flour cup with a knife. At first I check every ingredient before they mix it with anything else.

The utensils needed are:

- A large bowl for dry ingredients
- A medium-sized bowl for the liquid and fat ingredients
- A portable electric mixer that can be used in the bowls
- A square baking pan
- A rectangular baking pan
- A set of measuring cups with the fractions labeled clearly on each cup (for each student, if possible). You may be able to find different colors for each cup or cups with color-coded fractions.
- A set of measuring spoons with the fractions labeled clearly (for each student, if possible) The labels on teaspoons tend to be quite small, so it would be good to have a set where the labels or spoons are different colors.
- Mixing spoons

- Oven timer
- Oven or microwave

I also cover the table with newspaper and put used packages and trash into a small plastic bag to be disposed of later.

I review the different cup measures and the measuring spoons before each session. The older students read the ingredients out loud and show which measuring cups or spoons they need to use. For the younger students, at first we go through each measurement and step as a group. Later we just check vocabulary before they began.

The students should have time to put away the ingredients, clean the table, and wash the cooking dishes after each session. Of course, they should be able to eat some of the results at the end of the class. The first recipes listed (for cookies or waffles) have a short baking time so students can eat them at the end of each session. The more useful recipes such as muffins and breads that are listed later have longer baking periods such as 45 minutes. At home, this will probably be no problem, but teachers at school may have to arrange to have the students come back at a later time to sample the results.

Teachers have told me that they have used a microwave successfully for some of these recipes. The microwave is faster, and more affordable for schools. I have found that microwaves vary in temperature. Toward the end of the cooking time, check your baking goods frequently. If your microwave does not have a turntable, rotate your pan once in the middle of the cooking cycle.

Students always seem to enjoy the baking sessions—and the eating. I have found that some older students already knew how to make simple lunches or dinners that don't require much measurement. Baking is often a new experience for them, however. Students I have worked with needed quite a bit of supervision in the beginning. Later they went right to the tasks and did them correctly—with an occasional lapse. They often asked to do a certain recipe a second time and could usually do the task better than the first time. There was never any question about why fractions were needed because the students knew why from practical experience.

If you are working with younger students, you may have to do some of the preparations yourself. The purpose of baking is to show why fractions are useful and associate a pleasant experience with them. If the baking itself is too difficult for your students, you will defeat the purpose of the activity. Have the students measure the ingredients each time. You may have to crack the eggs or do the final mixing or handle the oven, but they will still learn about fractions and have fun eating the results.

BAKING USING ½ CUP MEASURES

PINEAPPLE POUND CAKE

OBJECTIVE: The student will be able to measure ½ cup of ingredients to use in baking.

INGREDIENTS:
- 1 package yellow cake mix
- 1 package vanilla pudding mix (instant)
- 4 large eggs
- 1 can pineapple slices (or crushed pineapple) drained
- ½ stick of butter
- ½ cup packed brown sugar
- ½ cup pineapple juice, drained from the canned pineapple
- ½ cup water
- ½ cup oil
- Angel food or bundt cake pan or two pie-sized small cake pans
- Cooking spray

DIRECTIONS:
1. Combine cake mix, pudding, eggs, oil, water, and pineapple juice in large bowl.
   Put in: a. 1 package yellow cake mix
   b. 1 package vanilla pudding
   c. 4 eggs
   d. ½ cup oil
   e. ½ cup water
   f. ½ cup pineapple juice
2. Beat for 2 minutes with electric mixer.
3. Melt ½ stick butter in microwave. (Be sure to have a student cut the stick in half.)
4. Stir in ½ cup brown sugar.
5. Spray pan thoroughly with cooking spray or shortening.
6. Arrange pineapple slices or crushed pineapple on the bottom of pan.
7. Put brown sugar and butter mixture in small pieces over the pineapple.
8. Pour cake batter over pineapple slices.
9. Bake at 350 degrees for 50-60 minutes.
10. Let cool and then invert the pan over cake plate and serve.

MOIST SUGAR COOKIES

Now that the students have used the ½-cup measures in the previous recipe, you may want to introduce ½-teaspoon measures. Show the students the 1 cup measuring cup compared to the ½-cup measuring cup. Then show the 1 teaspoon as compared to the ½ teaspoon. Have the student use two ½-teaspoons of rice to fill the 1-teaspoon measure. If you feel that the student understands the use of ½ cup versus ½ teaspoon, have him bake a recipe that has both ½ cups and ½ teaspoons in it, as follows:

OBJECTIVE: The student will be able to use ½-cup and ½-teaspoon measures in baking.

INGREDIENTS:

- ½ cup butter or margarine, softened
- ½ cup sugar
- 1 egg
- ½ cup cooking oil
- ½ cup powdered sugar
- ½ teaspoon cream of tarter
- ½ teaspoon baking soda
- 2 cups flour

DIRECTIONS:

1. In a large bowl, mix the ½ cup butter and ½ cup regular sugar. Stir until smooth.
2. Add 1 egg and ½ cup cooking oil and stir.
3. In a medium bowl, mix together:
   - a. ½ cup powdered sugar,
   - b. ½ teaspoon cream of tartar,
   - c. ½ teaspoon baking soda. Stir.
4. Add dry ingredients to butter, egg, and oil mixture and stir well.
5. Drop mixture by teaspoonfuls onto a cookie sheet.
6. Press the cookies down with the bottom of a glass.
7. Bake at 350 degrees for 7 minutes. Do not over bake.

Scott was measuring the dry ingredients for waffles into a bowl. He started to pour salt into a measuring cup. An assistant called to him, "Scott, you are going to have awfully salty waffles if you use that much salt." Scott replied, "But the recipe says to use ¼ cup salt." "Look again, Scott," said the assistant, "the recipe says ¼ TEASPOON salt." The assistant then had Scott read aloud the complete fraction, including the name of the unit (teaspoon or cup), before actually measuring the ingredients.

BAKING WITH HALVES AND FOURTHS

MOIST OATMEAL COOKIES

OBJECTIVE: The student will be able to use halves and fourths of both cups and teaspoons in simple baking recipes.

INGREDIENTS:

- ¾ cup shortening
- 1 cup brown sugar
- ½ cup sugar
- ¼ cup water
- 1 egg
- 1 teaspoon vanilla
- 1 cup flour
- ½ teaspoon baking soda
- 1 teaspoon salt
- 3 cups quick-cooking oatmeal
- 1 cup chopped nuts
- 1 cup raisins

DIRECTIONS:

1. In a large bowl, mix together the following indredients until fluffy and smooth:
   - ¾ cup shortening
   - 1 cup brown sugar
   - ½ cup white sugar

- ¼ cup water
- 1 egg
- 1 teaspoon vanilla
2. In a medium bowl, mix together
   - 1 cup flour,
   - ½ teaspoon baking soda
   - 1 teaspoon salt
3. Add these dry ingredients to the mixture in the large bowl.
4. Stir in oatmeal until well blended.
5. Add nuts and raisins (optional) to the mixture.
6. Drop by teaspoonfuls onto greased cookie sheets.
7. Bake at 350 degree for 12-15 minutes or until lightly golden.

MICROWAVE APPLESAUCE CAKE

The following recipe using halves and fourths can be baked in a microwave:

INGREDIENTS:
- 2 cups flour
- ¾ teaspoon baking soda
- 1 teaspoon cinnamon
- ½ cup raisins
- ¼ teaspoon salt
- 1 cup sweetened applesauce (can be chunky)
- ¼ cup butter or margarine
- 1 cup brown sugar, packed
- ½ teaspoon cloves
- ¼ teaspoon nutmeg
- Canned frosting, optional
- Cooking spray

DIRECTIONS:
1. Spray an 8" microwavable (glass, silicone, or heavy plastic) pan with nonstick spray.
2. Put butter in microwave in a large microwavable bowl for 40 seconds until melted. (Again, remember to have a student cut off ¼ of the stick.)
3. Add 1 cup brown sugar and stir.
4. Add 1 cup applesauce and blend well by hand.
5. In another bowl, mix:
   a. 2 cups flour
   b. ¾ teaspoon baking soda (show the students how to fill the ¼ tsp 3 times)
   c. 1 teaspoon cinnamon
   d. ½ teaspoon cloves
   e. ¼ teaspoon nutmeg
   f. ¼ teaspoon salt

Note: microwave ovens vary in temperature. Check the cake with a toothpick after the first 6 minutes.

6. Add the dry ingredients to the sugar, butter, and applesauce mixture.
7. Stir thoroughly—perhaps with electric beaters.
8. Add ½ cup raisins, stirring with a spoon.
9. Spread mixture in microwavable pan.
10. Microwave on high for 6 minutes. Rotate and cook for 2 more minutes, checking to see if done after each minute.
11. Cool in pan and frost if desired.

THIRDS

The other commonly used fractions are thirds. Recipes often call for ⅓ and ⅔ cups. Thirds, however, cannot be easily converted into halves, fourths, or eighths. Teach thirds as separate fractions without reference to the halves and fourths that the students have already learned.

INTRODUCING THIRDS

OBJECTIVE: The student will be able to read the fractions ⅓ and ⅔ and demonstrate their meanings.

MATERIALS:
- Strips of construction paper divided by lines into thirds
- ⅓ measuring cup
- 1 cup of rice
- Clear bowl
- Slate, whiteboard, or pads of paper
- Marker or chalk

PROCEDURE:
1. Using the construction paper strips divided into thirds, teach the students about thirds in the same manner as you taught about fourths above.
2. Have the students make ⅓ and ⅔ out of their construction paper strips by tearing or cutting them and test each other.
3. Have one student measure ⅓ cup of rice with the measuring cup and pour it into a clear bowl. Then have another student pour a second ⅓ cup of rice into the bowl. Ask the students to tell how much rice is now in the bowl (⅔ cup). Have them measure with the measuring cup to check.
4. Have the students write the fractions ⅓ and ⅔ on a slate, whiteboard, or pad of paper. Repeat procedures until you think the students are secure. Also review ½, ¼, and ¾ cups using the rice and cups.

BAKING WITH THIRDS

CHEWY CHOCOLATE BROWNIES

OBJECTIVE: The student will be able to use thirds in baking.

INGREDIENTS:
- 1 large package regular chocolate pudding mix (Do not use instant pudding)
- 3 eggs
- 2 teaspoons vanilla
- 1 cup chopped nuts (optional)
- ⅓ cup sugar*
- 1 cup flour
- ½ teaspoon baking powder
- ⅔ cup butter or margarine
- 1 cup sugar*

*The recipe actually calls for 1⅓ cup sugar, but we haven't introduced mixed numbers yet so the sugar is presented as two separate ingredients.

DIRECTIONS:
1. In a large bowl, stir together until blended:
   - a. pudding mix
   - b. 1 cup flour
   - c. ½ teaspoon baking powder
2. Melt ⅔ cup butter (in microwave or on the stovetop) and put in another bowl.
3. Stir in 1 cup sugar. Then stir in ⅓ cup sugar.
4. Beat in 3 eggs, one at a time.
5. Blend the dry ingredients with the butter, sugar, and egg mixture.
6. Add 2 teaspoons vanilla to the mixture and stir.
7. Stir in nuts (if desired).
8. Spread in a greased 9 x 12 pan.
9. Bake at 350 degrees for 20-25 minutes. Do not over bake.

MICROWAVE BROWNIES

OBJECTIVE: The following recipe uses thirds and can be made in a microwave.

INGREDIENTS:
- 1 cup flour
- ⅓ cup unsweetened cocoa
- ¼ teaspoon baking powder
- ¼ teaspoon salt
- ¾ cup chopped nuts (optional)
- cooking spray
- ⅔ cup butter or margarine
- ⅔ cup white sugar
- ⅓ cup brown sugar, packed down
- 2 large eggs
- 1 teaspoon vanilla extract

DIRECTIONS:
1. Coat an 8 or 9 inch microwavable square or pie pan with nonstick spray.
2. In a small bowl, microwave butter 20 seconds or until softened and pour into a large bowl.
3. Measure ⅔ cup white sugar and ⅓ cup brown sugar. Stir into the butter.
4. Add 2 eggs and 1 teaspoon vanilla and beat until well blended.
5. In another bowl, mix:
   - a. 1 cup of flour,
   - b. ⅓ cup cocoa powder,
   - c. ¼ teaspoon baking powder
   - d. ¼ teaspoon salt.
6. Add the dry ingredients slowly to the egg and sugar mixture. The dough will be very thick.
7. Stir and then add nuts (optional).
8. Spread the brownies evenly in the microwavable pan.
9. Microwave at 100% power for 5 to 7 minutes. After 5 minutes, check frequently (and rotate if needed).
10. Take the brownies out of the microwave. Set pan on a rack or a small bowl overturned to cool.
11. Sprinkle with powdered sugar or frost lightly if desired.

VELVET WAFFLES

OBJECTIVE: The following recipe uses thirds and can be made in a waffle iron. Makes 4-5 waffles.

INGREDIENTS:

- 1 cup flour
- 1 tablespoon baking powder
- ½ teaspoon salt
- 1 tablespoon sugar
- 2 eggs
- ⅔ cup milk
- ⅓ cup oil
- Cooking spray

Note: This is a favorite recipe of the young adults I have worked with. It requires a waffle iron and a little guidance during the baking process. They especially like it with strawberries and a little whipped cream.

DIRECTIONS:

1. Heat waffle iron and spray with cooking spray.
2. Mix dry ingredients together in large bowl.
    a. 1 cup flour
    b. 1 tablespoon baking powder
    c. ½ teaspoon salt
    d. 1 tablespoon sugar
3. Crack 2 eggs into small bowl and beat them with a fork.
4. Add ⅓ cup oil and ⅔ cup milk to the eggs in the small bowl.
5. Mix the liquids into the dry ingredients in the large bowl and stir or beat with electric beaters until smooth.
6. Use a half-cup measure to pour waffle batter onto waffle iron.
7. Cook the waffles, without opening the lid, until no steam is visible.

Other recipes are available in Appendix C. The students I worked with seemed to learn very much about fractions and baking skills while making these recipes. They enjoyed the cooking so much (and the eating) that we did all the recipes in this chapter and in Appendix C.

Other Common Fractions

The fifths and sixths are not that commonly encountered in everyday life, but by naming them (as well as other fractions such as eighths and sixteenths) the student can show that he understands the pattern used in naming all fractions. Note that fractions such as ³⁄₆ that can be reduced to smaller terms are not taught here.

INTRODUCING FIFTHS, SIXTHS, AND OTHERS

OBJECTIVE: The student will be able to read the common fractions (½, ⅓, ¼, ⅕, ⅙, ⅔, ¾, ⅖, ⅗, ⅘, and ⅚) correctly.

MATERIALS:

- Chalk and slate or whiteboard and marker

- Two copies of the Fraction Strips sheet (from Appendix B). (If possible, one sheet should be copied onto cardstock and laminated before cutting for further use. The second sheet is for the student to look at when he is working with the fraction strips.)
- Pen, pencil, or fine-line marker
- 6 different small objects such as a pencil, eraser, etc.
- Egg carton cut in half—leaving 6 cups
- 6 poker chips or magnetic bingo chips

SUCCESS STEP: Have the students arrange six objects in line. Say, "The pencil is first in line, the eraser is second. What position is the tennis ball in? (third).

PROCEDURE:

1. Tell the students that when naming fractions, you use the same words that you use for telling a position in line—third, fourth, fifth, sixth, seventh, eighth, etc.
2. Write fractions using 1 on the top and 2 through 6 on the bottom. Help the students to read them correctly. Tell them that ½ is the naughty one. He is not called 1 twoth or 1 second, but one half (maybe twoth sounds too much like tooth).
3. Review naming fractions for a few minutes at the beginning of several sessions. When the students are secure with the naming of fractions with 1 on the top, introduce fractions such as ¾ with other small numbers on top.
4. Show the egg carton with the 6 cups. Have the students put 1, and then 2, etc. poker chips in the cups. Give each person a turn and have everyone call out the name of the fraction that is formed; e.g., ⅙, ⅖, etc.
5. Have the students cut apart the top half (1—1⁄16's) of the fraction strip sheet from Appendix B.

Fraction Strips

6. Have them label the 1, ½'s, ¼'s, ⅛'s, and the 1/16's with a pen or marker. Have them color each line a different color.

7. Cut out first the 1 whole, then the halves, etc., with you cutting your own sheet as the students cut theirs. Have them put fractions (each a different color) in separate piles in front of them.

8. Practice with the cutout pieces, showing how 2 halves make 1 whole and 4 fourths also make one whole, etc. Lay the cut-out pieces on the 1 whole to help the students see this relationship.

9. Call out fractions such as one whole, one-half, three-fourths, two-eighths, or three-sixteenths. Have the student show you that fraction with his fraction strips. Practice on other occasions.

10. When the student is confident with the fractions from the top half of the worksheet, continue as above using the tenths and the fifths.

11. Then proceed as above using the thirds, sixths, and twelfths. These fractions are probably the most commonly used ones in daily life.

Teaching the Parts of a Fraction

It may be important for the students to know the names of the parts of the fraction, especially if they are working in the general education classroom. The numerator is the number on the top of the fraction or the one that is written first. The denominator is the number on the bottom or the one that is written last. If appropriate, you can tell them that if you take the first two letters of the numerator and the first two letters of the denominator, you have NU DE (NUDE!). Using this memory aid, you can tell which is the first number written (NU) and the second number written (DE).

FRACTIONS AS WHOLE NUMBERS

OBJECTIVE: The student will demonstrate that when a fraction has the same number on the top and the bottom of the fraction, the fraction is really the whole number 1.

MATERIALS:
- ¼ cup, ½ cup, ⅓ cup, and 1 cup measuring cups
- 2 cups rice
- 2+ cups of water (colored with a little food coloring)
- Slate, whiteboard, or pad of paper with marker or chalk
- Fraction strips from previous activity

SUCCESS STEP: Ask the student to name the ¼-cup and ½-cup measures. Praise him for the correct answers or point out the correct cups and have him repeat the words.

PROCEDURE:
1. Have the students put 4 one-fourth cups of rice in the 1-cup measure. Write the fraction 4/4 on the board. Show them that whenever the fraction has the same top and bottom number it means that one whole unit is there. Write 4/4 = 1 on the board.

2. Then write ⅖ on the board. Ask each student to put 2 one-half cups of rice in the cup measure and identify it as 1 whole unit.
3. If you have ⅓ cup measures, ask the students how many thirds of a cup make one whole (3).
4. Ask them to chant:

   Fractions with the same
   bottom and top
   Are really number 1
   so you can stop.

5. Have the students repeat steps 1 and 2 using water instead of rice. It is easier for them to see the quantity if you color the water slightly (and more fun also).
6. Using the fraction strips from the previous activity, have the student show you 2 two's, 3 three's, 4 four's, 5 fives, etc. Compare them to the 1 whole strip. Emphasize that these fractions are equal to one whole.

GENERALIZATION ACTIVITY: The next time you make a recipe for cake, brownies, etc., cut it into 6ths or 8ths or another fraction. Ask the student(s) to tell you what fraction you've cut it into (e.g., 6ths). Before you give out any of the pieces, point out that the whole has now been cut into 6/6 or 8/8. Ask them if that is the same as one. Do the same when you are cutting sandwiches or pizza at home.

DOMINO FRACTION GAME

OBJECTIVE: The student will be able to demonstrate by playing the game that he can read fractions correctly, including fractions which are equal to 1.

Domino Fractions-2

/6's	/2's & /3's	/4's	/5's
3/6	2/3		4/5
		1/4	1
2/6	1		
5/6	1/2	1/4	3/5

MATERIALS:

- Dominoes or domino fraction cards. (Real dominoes are preferred for this game but domino fraction cards are also found in Appendix B. If using real dominoes, use only the dominoes with two sets of dots on them; put away the dominoes with only one set of dots.)
- Individual domino fraction boards for each student, as found in Appendix B

PROCEDURE:

1. Show some of the dominoes to the students, holding them so the half with the most dots is on the bottom. Tell them that the dots on the top of the domino represent the top number of the fraction, the line in the middle of the domino represents the line/slash of the fraction and the bottom section of the dots is the bottom of the fraction for this game. Turn all the dominos over on the table so the dots do not show.
2. One student turns over a domino, making sure that the smallest number of dots is on the top. He then names the fraction that it represents. If he has that fraction on his

game board, he puts the domino on top of it. If he doesn't have the fraction, he turns it over again on the table and another person takes a turn.

3. If a student turns over a fraction that is equal to 1, such as ⅓, he may put it on the number 1 on the game board.

4. The first student to cover one line on the game board—up or across—is the winner. Alert the students to watch where the dominoes are that have not been used.

5. The important part of this game is to make sure the students are naming the fractions correctly. An adult or older student should play with a beginner to make sure he names fractions correctly. The game boards do not have any fractions that have the same number on the top as on the bottom. If the student picks up any domino that has the same number on the top and on the bottom, he should recognize that it is equal to 1. There are several 1s on each fraction board, so he can use that domino to cover any of those 1s.

6. The fractions are arranged on the fraction boards with one column showing sixths, another showing halves and thirds, another fourths, and another with fifths. Alert the students to this pattern so they can search their cards more efficiently.

7. The Domino Fraction Faction game is played quite quickly like a combination of tic-tac-toe and Bingo. Students should be able to play several games at a time and keep track of their winnings.

Note: If you think that the students are having trouble with the procedure of counting the dots and then reading the numerals of the fractions, make a set of cards with the numerals instead of the dots and have the students read the fractions.

Lowest Terms

It is not necessary to teach your student(s) how to reduce *all* fractions to their lowest terms at this point. For now, teach them that ²⁄₄ and ⁴⁄₈ are the same as one half. This will familiarize them with the idea that some fractions are equivalent to each other. On a practical level, it will also help them see that it really doesn't make any difference if they use 2 fourth-cup measures or 1 half-cup measure while baking—they will get the same results. Just show them that the amounts are the same so some recipes will not confuse them.

INTRODUCING FRACTIONS EQUIVALENT TO ONE HALF

OBJECTIVE: The student will be able to show that ½, ²⁄₄ (and ⁴⁄₈) are the same amounts.

MATERIALS:
- ¼-cup and ½-cup measuring cups for each student
- (⅛-cup measuring cup, if you have one)
- Approximately one cup uncooked rice or sand for each student
- Slate or whiteboard and chalk or marker
- Newspapers to catch spills

SUCCESS STEP: Have the student identify the ¼-cup and the ½-cup measures. Praise him if correct or show him what each one is and then ask him to identify them again. He may look at the labels on the measures.

PROCEDURE:
1. Ask the student which of the two measuring cups is larger (the half-cup measure). Then ask the student to fill up the ¼-cup measure and dump the rice into the ½-cup measure. Then ask him to put another ¼ cup of rice into the ½-cup measure. If he is careful, the ½-cup measure should now be full.
2. Write on a slate or whiteboard the sentence: ¼ cup rice and ¼ cup rice = ²⁄₄ cup rice, which is the same as ½ cup rice. Then write directly below that sentence:
   ¼ + ¼ = ²⁄₄ or ½.
3. Have each student repeat step 1 himself and experience that two fourths equals one half.
4. If you have ⅛-cup measuring cups, repeat steps 1 and 2, pouring four ⅛ cups into the half-cup measure and writing:
   ⅛ cup + ⅛ cup + ⅛ cup + ⅛ cup rice = ⁴⁄₈ cups rice or ½ cup rice.
5. Review the above activities several times at different math sessions. (You may wish to change the materials to cereal or small dried beans so that the students can generalize to other materials.)
6. Ask each student, "Which is larger, ²⁄₄ cup of rice or ½ cup of rice?" They should say that they are the same.

Mixed Numbers

Students often have difficulty reading and writing mixed numbers—numbers that have a whole number and a fraction. Perhaps the small difference in spacing between the whole number and the numerals in the fractions make it difficult to see which is the large number. When the fraction is written as numeral /(slash) numeral (e.g., 5/6), there is no difference in size to use as a cue. Sometimes the students see the whole number as part of the fraction, such as reading 2 5/6 as 25 sixths. When first introducing mixed numbers, I put two or three spaces between the whole number and the fraction to make the difference more obvious (2   5/6). Later, the extra space can be faded so the students can read mixed numbers in commercial textbooks, workbooks, etc.

INTRODUCING THE CONCEPT

OBJECTIVE: The student will be able to correctly read mixed numbers (1- to 2-digit whole numbers and simple fractions) and demonstrate understanding with real objects.

MATERIALS:
- Construction paper with 5 to 6 pieces of the same color for each student
- Teacher-made sets of whole and fractional sheets of construction paper to represent the mixed numbers: 1½, 2½, 3½, 1¼, 1¾, 2¾, 3¾. Make each set using the same color of paper (e.g., represent 1½ by putting a whole sheet and a half sheet of blue construction paper together.)
- Various fruits (apples, oranges, pears, etc.) as used above, one fruit being cut in half
- Candy or granola bars

- Knife for teacher
- Slate, whiteboard, or pad of paper and marker or chalk

SUCCESS STEP: Ask the student to give you ½ of one of the fruits. Praise him for the effort.

PROCEDURE:

1. Hold up the ½ fruit and then a whole one of the same fruit. Ask the students how many of the fruit you have in your hands. Help them to see that you have 1 whole fruit and ½ fruit.
2. Ask, "How do you think we can write this number?" Using the board or pad, write the number 1½. Point out which is the whole number and which is the fraction.
3. Give each student 1 and ½ pieces of the same color construction paper. Ask them how much construction paper they now have. They should say, "One and one-half pieces of paper." Point out that you say the word *and* between the whole number and the fraction. If they say, "1 big piece and 1 little piece," say, "This is a one WHOLE piece and this is HALF of a piece."
4. Give the students one more piece of the same color construction paper. Have them tell you that they have two and one-half pieces. Let each student write *2 and ½* on a paper for himself. Then have him write 2 ½ as is normally done (but with a little extra space between the whole number and fraction).
5. Write mixed numbers on the board and see if the students can name and make them with construction paper. Use mixed numbers such as 3½, 1½, 2¼, 2¾, 3¼ with small whole numbers and only the fractions that the students already know.
6. Have the students make their own mixed numbers to show to you or the rest of the class. Remind them that they can fold and tear the pieces of construction paper into halves and fourths. Have one student show the pieces of construction paper that make up his mixed number and see if the others can read it. If you have only one student, you will have to read the student's mixed number yourself.

IMPROPER FRACTIONS (OPTIONAL)

OBJECTIVE: The student will demonstrate understanding of what an improper fraction is by using manipulatives and be able to convert it to a mixed number.

MATERIALS:

- Same as previous activity

PROCEDURE:

1. Using the construction paper pieces from the above activity, make a simple mixed number such as 1½.
2. Ask the students how many fraction pieces (halves) the whole could be divided into (2).
3. Draw a line down the middle of the whole sheet to show the two halves. Touch the halves on the whole sheet and the half sheet as you count aloud (1, 2, 3 halves). Write the mixed number as an improper fraction, so 1½ becomes ³⁄₂.

4. Tell them that the fraction is top heavy because the 3 is larger than the 2. Draw a picture of the top-heavy fraction ³⁄₂.

5. Say that most of the time we need to make the improper (top-heavy) fraction into a mixed number. Show the mixed number of 1½. Use the mixed numbers that the students made or the mixed numbers that you have been teaching with to illustrate what to do with top-heavy fractions. Write or draw on the board both the improper fraction and the mixed number that it becomes. Remind them that a fraction is always smaller than one unit (whole object), so we have to see how many whole numbers can be made out of the top-heavy fraction and then write the fraction left (if any). You may need to remind them that any fraction that has the same top number as bottom number equals 1 whole. (See domino game above.)

6. If you want to show them a quick way to get the mixed fraction, you can tell them to divide the bottom number into the top number. Only do this when they understand from working with the construction paper manipulatives why that will work. To use this method, it helps to have one of the calculators mentioned in the division chapter that can be programmed to show fractions rather than decimals.

Fraction Equivalents

Commercial materials often use pies or pizzas to demonstrate fractions. I have found that dividing pies works well for halves, thirds, and fourths. However, when the circles are divided into fifths, sixths, or more pieces, it is difficult to see how the slices are different, especially if the fractions are close in value. You may want to use commercial learning games such as *Pizza Party* and *Pizza to Go* (Ideal) or *Pie in the Sky* (Learning Resources) for their interest value, but do not be discouraged if the student has difficulty with the smaller fractions. (See Resources for sources of commercial games.)

It might be important for your student to learn to do the typical type of worksheet often given in schools for fractions. The Fraction Strips chart found in the Appendix shows one whole (8½ inches long) divided into halves, fourths, eights, sixteenths, and also thirds, fifths, sixths, tenths, and twelfths. This worksheet is semi-concrete, however. On this worksheet, the fractions, similar to the construction paper pieces the students have been using, are now turned into strips that make it a little easier to see equivalents when they are divided into many small fractions.

FRACTION STRIPS EQUIVALENTS

OBJECTIVE: The student will be able to demonstrate with fraction strips what fractions are equivalent to each other.

MATERIALS
- Fraction strips used in the section "Other Common Fractions" above
- Uncut Fraction Strip sheet
- Fraction Strip Worksheet from Appendix B
- Chalkboard or slate, chalk
- Equivalent Fractions Worksheet (Appendix B)

PROCEDURE:

1. Go through the first 5 fraction strips (1, 2, 4, 8, and 16) and point out what the equivalents are. Have the students label each division with black pen or markers if not already done in the section "Other Common Fractions," above.
2. Have the students put the cut fraction pieces on the uncut Fraction Strip sheet to show that 4 eighths are equal to 1 half, 2 fourths are equal to 4 eighths and 1 half, etc.
3. Then do the same with the tenths and fifths.
4. Later, do the fraction strips that show the equivalents of thirds, sixths, and twelfths
5. Using the slate or chalkboard, show the students that 4 of the 16ths can be written ⁴⁄₁₆.
6. Call out some of the fractions on the top part of the fraction chart and have the students point to them on their strips. When you think that they have the idea, do two problems on the Fraction Strip Worksheet with them and check their answers.
7. Have the students independently do the Fraction Strip Worksheet. They need only recognize the divisions of the fractions to do this worksheet. Repeat if necessary.

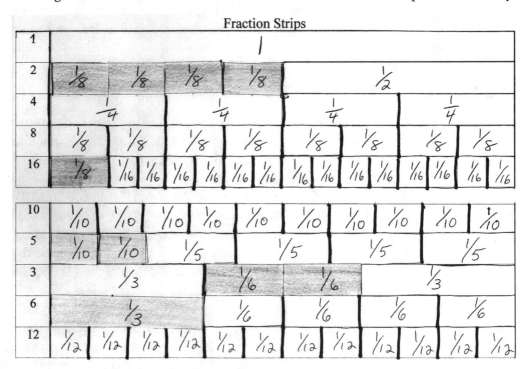

Fraction Strips

Adding Simple Fractions

In survival math, there is not much need to add or subtract fractions, especially fractions with different denominators. True, adults who are doing some or all of their menu planning or shopping independently may sometimes encounter situations where being able to add fractions might be helpful. However, there are usually practical ways to get around adding the fractions. For example, an adult might want to make two batches of a muffin recipe. If he knew how to add fractions, he would realize that if one batch calls for ⅔ cups of sugar, two batches must call for ⁴⁄₃ or 1⅓ cups. If he cannot add the fractions, however, he can measure out ⅔ cups of sugar twice and get the proper amount. For another example, he may be planning to make recipes during the week that call for ½ cup of milk, 2¼ cups of milk, and 1½ cups of milk. Some of us might add up the amounts and make sure we have at least 4¼ cups of milk on hand. But others of us (even those of us who can add fractions) might just buy a half gallon or gallon and assume that will be plenty.

However, especially if the students are participating in the general math curriculum, it is important to learn how to add fractions with the same denominator (with the same last names). It is also important to understand that you cannot add fractions together if their denominators are *not* the same (for instance, adding ½ and ¼ to get 2/6).

INTRODUCING THE CONCEPT

OBJECTIVE: The student will be able to add fourths together. Later he will be able to add thirds.

MATERIALS:

- Slate or whiteboard and chalk or marker
- Juice, water, or other beverage
- ¼-cup measuring cup
- Clear drinking glasses

SUCCESS STEP: Ask the student to pour the juice or other beverage into the ¼ measuring cup. Assist him, if necessary, and praise.

PROCEDURE:

1. Ask the students to pretend that they have been running and are very thirsty. "Would this much juice (water) be enough for you to drink?" (No)
2. Tell the students you want them to show you how much they would want to drink when they were thirsty, but that you want to measure the total amount they pour into the glass. How could you measure the amount if you only have a ¼ measuring cup? They should say to use the ¼ measuring cup several times.
3. Let one student pour the ¼ cup of juice into the clear drinking glass. Then ask them how many ¼ cups to pour in the drinking glass and do the pouring.
4. Write the fractions (¼) on the board according to the number of quarter cups that have been poured in the drinking glass. Show the students how to add the top numbers together, e.g. ¼ + ¼ + ¼ = ¾. Don't let them pour more than 4 of the ¼ cups into the drinking glass.
5. Tell the students, "You can only add the top numbers if the bottom numbers are the same.... If they have the same last names (fourths)."
6. Have one student pour a beverage by ¼ cups into the drinking glasses while the teacher or another student writes the number sentence on the slate.
7. Let the students drink the beverage.
8. Later have the students repeat the procedure using the ⅓ cup measure.

Note: *You may want to show the students what happens if the bottom numbers (denominators) are not the same. Show them that ½ + ½ cups of drink fills the whole cup. BUT if you add ½ + ½ WRONG by adding the top numbers and the bottom numbers you get the fraction 2/4 which is a lot smaller than what the real drink shows.*

Common Denominators

Sometimes the student may be included in general math classes or need to add or subtract fractions with different denominators. Math classes usually spend a lot of time learning to find the lowest common denominator so fractions can be added or subtracted. Most of the time the fractions involved have one-digit denominators. Why not use the easiest common denominator for two fractions—the product (answer in multiplication) of the denominators of each fraction? For example:

$$\tfrac{1}{6} + \tfrac{1}{4} = \underline{\qquad}.$$

Conventional math requires you to find the lowest common denominator, which is 12. Then you have to figure out how many 12ths are the same as ⅙ and ¼. Then you add $\tfrac{2}{12} + \tfrac{3}{12} = \tfrac{5}{12}$.

The easiest common denominator would be $6 \times 4 = 24$ (multiply denominators by each other).

It is easier to convert ⅙ to 24s by multiplying by the other denominator (4). $\tfrac{1}{6} = \tfrac{4}{24}$. Likewise $\tfrac{1}{4} = \tfrac{6}{24}$. $\tfrac{6}{24} + \tfrac{4}{24} = \tfrac{10}{24}$, which is the same as $\tfrac{5}{12}$. (You will probably have to teach the student to reduce to lowest terms. Again, a fraction calculator can be handy for these kinds of operations.)

GENERALIZATION ACTIVITIES

- Practice folding a letter into thirds so it will fit in a business-size envelope.
- Find a recipe for potpourri that uses fractional amounts of ingredients. Help your child make it as a present for someone.
- Make homemade play dough using a recipe that uses fractional amounts. You can find many interesting recipes for play dough on the Internet such as Peppermint Play Dough, Jell-o Play Dough, and Edible Play Dough. For examples of recipes, see www.recipegoldmine.com.
- Make a special mixture of birdseed for your feeder using fractional amounts of different kinds of seeds (e.g., 1½ cups of sunflower seed, ¾ cups of millet seed, 1⅓ cups corn kernels).
- Make your own trail mix, designing the recipe so your student uses the fractions he needs to practice (for example, 1½ cups of cereal, ¼ cup of M&M's or chocolate chips, ⅔ cup of raisins, ⅓ cup peanuts).

- Divide up chores at home. Tell your child that he needs to vacuum/sweep half the floor (or shovel half the driveway or rake half the yard) and another family member (or you) will do the other half. Agree on where the halfway mark is. At school, ask one student to clean half the blackboard or pick up papers from one half of the class, while another student does the other half.
- When you are trying out a new cereal (or other food item such as chips, peanuts, raisins, or ice cream measured in cups), check the packaging to see what a serving size is and then measure out that amount.
- If you are watching your cat's or dog's weight, let your child measure out the right amount of dry food for your pet.
- Make some of the recipes included in Appendix C. Or search the Internet or cookbooks for appealing recipes that include fractions and that your child or student can help you make or can make independently.

Commercial Games to Buy or Try

- Play **One! The Fraction Solution** (card game, available from educationallearninggames.com). There are several games to play using cards with visual representations of the most common fractions. The games are designed to teach about fractions that equal one, as well as equivalent fractions.

- Play **Pie in the Sky** (www.learningresources.com). You spin to see what fraction of a pie you earn and place the pieces on pies on a board, trying to build a whole pie.

CHAPTER 12

Measurement

Questions to be answered:

Can the student:

1. Measure length and width to the quarter inch.

2. Compare the familiar and metric systems of measurement as far as names of the measures and what they measure.

3. Compare the perimeter and the area of a rectangle.

4. Relate miles to walking and riding in a car.

5. Identify cups, pints, quarts, half gallons, and gallons and be able to demonstrate their equivalents.

6. Read the temperature on a large thermometer.

7. Demonstrate understanding of telling time by *minutes before and minutes after*.

8. Demonstrate understanding of *quarter before* and *quarter after* using a clock.

9. Demonstrate the meaning of time vocabulary such as *ago, half past, quarter past, quarter 'til, almost, in ____ minutes* during daily living activities.

In *Teaching Math, Book 1,* the concept of measuring length by self and standard units was explored along with some basic information on weight, capacity, and temperature. In the section on length, we discussed using rulers with inches, feet, and yards. Students who have completed the first book should have had lots of experiences measuring things in their environment using increasingly complex rulers with markers at the quarter inch (or smaller).

In this volume of *Teaching Math,* rulers and tape measures will be used not only to measure length, but also to teach some concepts about perimeter and area of simple shapes. We will also build on the measurement of capacity introduced in the Fractions chapter, where we used teaspoons, tablespoons, and cups to measure. In this chapter we will explore pints, quarts, half gallons, and gallons. In the topic of weight, the relationship of pounds, grams, and ounces will be discussed.

If you live in an area where the metric system is in common use, you can spend much less time working on the equivalents of one measure to another because everything is based on 10. If possible, students in the USA should have at least some exposure to the metric system. At a minimum, they should learn to recognize metric terms such as meter, centimeter, liter, grams, and kilograms.

Reviewing How to Measure Inches and Feet

If the students have not completed the measurement chapter in Book 1 or if it has been awhile since they used a ruler, you may have to teach or re-teach them some basic concepts related to measurement. Basic concepts to make sure they understand include:

- Some rulers have only inches. Other rulers show inches on one side and centimeters on the other. If the sides are not labeled as either inches or centimeters, one clue is that the centimeter side goes up to 30 centimeters whereas the inches side goes up to only 12 inches. Centimeters are also much smaller than inches.
- How to place the edge of the ruler at the very edge of an item you are measuring and keep it straight. Ask your student to measure a few straight items and observe to see if she understands how to use the ruler.
- How to mark your place if you are measuring something that is longer than one foot with a ruler. See also if your student can replace the ruler correctly to count long edges. Students can also use yardsticks for measuring long edges.
- Math vocabulary with memory hints:
- **Length**—the **long** side of what is measured:
    ○ Length starts with **L,** as does Long.
- **Width**—The measurement from **side** to side or the smaller **side** of what is measured:
    ○ **W** is a **wide** letter (compared to other letters).
    Wide rhymes with **side.**
- **Height**—the distance from the base or bottom of something to the top, or how high something is:
    ○ Height starts with H, as does High.
- **Volume**—the amount of space that something occupies:
    ○ The letter **V** is shaped somewhat like a bottle or vase.
    ○ Fill up the **V** with a liquid or other stuff for volume.

MEASURING REAL OBJECTS

OBJECTIVE: The student will be able to measure length to the quarter inch.

MATERIALS:
- Rulers, if possible, ruled off in ¼ inches. (If ruler has markings down to 16th of an inch, highlight the ½ inch and ¼ inch marks using a marker—Make the ½ inch marks slightly longer than the ¼ inch marks—or use different colors)
- Yardsticks
- Slate or whiteboard

SUCCESS STEP: Ask the student to point to one inch on the ruler.

PROCEDURE:
1. Have the students use a foot-long ruler to measure the long end of a desk or the kitchen table and then the shorter side. If your table is round, find something else that is rectangular to measure.
2. Have the students measure to the ¼ inch or ½ inch, if the table is not an exact size. If they are not familiar with ¼ or ½ inches on the ruler, show them the ¼ and ½ inch marks. Have them record the numbers in total number of inches (not in feet and inches). Let them use a calculator to add up the inches, if necessary.
3. Have them measure the same desk or table with a yardstick. The result should be the same. Ask the students how many rulers make one yardstick? They should be able to show that three foot-long rulers equal one yardstick.
4. To help students remember that three feet equal one yard, ask them how many feet they have. When they say two, ask them, "What if you had three feet? Wouldn't you look funny with three feet? If you had *three feet,* I would send you out in the *yard.*"
5. Then have them measure some other things around the house/school with either the ruler or yardstick. Look for things to measure things that mean something to them personally. At home you may be able to have them measure for a rug or curtains, etc. that you actually want to buy.

Metric System

The metric system is used in most of the countries outside the USA. It is widely used in scientific communities everywhere. Although the superiority of this base-ten system has been acknowledged in the USA, long traditions have been established using the nonmetric familiar measures (the "English system") and we still continue to use them in our daily life.

Students in the U.S. will encounter the familiar measures in daily life much more frequently than the metric measures. However, students also need some experience with the names of the metric units, as they will encounter them in science class and also on some food products. They can use some of the metric system measurements without knowing the exact equivalents between the metric and familiar English units. For example, students should learn how to measure with a ruler, whether the units are inches or centimeters. For this purpose, the students should be aware of the most used metric units and the approximate familiar units. For example, they should have some idea that the units that measure lengths are either inches or centimeters.

UNITS OF LENGTH

OBJECTIVE: The student should be able to describe units that can measure length in the familiar (English) and metric system.

MATERIALS:
- Two pencils of equal length for each student
- A ruler that has both inches and centimeters on it for each student

SUCCESS STEP: Hold up a simple ruler (containing inches and centimeters) and ask the student what it is. Then ask her what it is used for. Praise success or model the answers for her and have her repeat them

PROCEDURE:
1. Put the ruler(s) in front of the students. Point out the inches scale and the centimeters.
2. Tell them, "We know about inches. Your mother and family use inches, the teachers in school use inches, even the principal uses inches. However, we are going to learn about another mysterious way to measure—centimeters. This way is the one that scientists use—the ones who build rockets and airplanes or who make the medicines we use. Look at the marks on this side of the ruler. It has marks that are closer together than on the inches side of the ruler. But both centimeters and inches can measure small things."
3. Remind them of the math vocabulary word *length* that they learned previously.
4. Have the students measure the one pencil with the centimeter side of the ruler. Then have them measure the other equal length pencil with the inches side of the ruler. Write the centimeters and inches on a whiteboard or slate. Underline the number of centimeters and label them as the mysterious metric measures that important people use.
5. Have the students measure at least 3 things in the room, using centimeters and inches and record the answers on a simple graph.

HOMEMADE RAIN GAUGE

Making the rain gauge will be a good activity for you and your student(s) if you live in an areas where there is enough rainfall to be noticed in a week or two, but that doesn't have big storms that will overfill a two-liter plastic bottle.

OBJECTIVE: The student will be able to measure depth to the ½ inch (later ¼ inch).

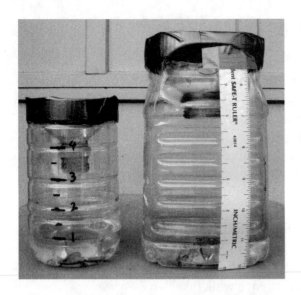

MATERIALS:
- One- to two-liter plastic water, soda, or juice bottle (with a flat bottom)
- Duct tape
- Small ruler

PROCEDURE:
1. The teacher or parent should cut the bottle about ⅓ of the way down so you have two units—one with the bottle top and the other with a big open bottle. Turn the part with the bottle top upside down and put it inside the larger bottle. This creates a funnel for the rain to flow though into the larger part of the bottle.
2. Have the students help put duct tape around the top edge to bind the two parts together.
3. Have the students tape a short ruler to the side of the bottle or mark the side of the bottle with inches and half (or quarter) inches using a permanent fine point marker.
4. Place the rain gauge outdoors in a place it won't be disturbed. You may need to put bricks or large rocks on several sides to keep it from being blown over in the wind.
5. Over the next few weeks, have the students measure the amount of rain each day and record the data on a chart. If there is little rain in your area, have the students wait a week before measuring it.

Larger Measures of Length

Students have probably heard the word *miles* in connection with driving or travel-ing. They do not have to have an exact knowledge of the length of a mile, but they should have some feeling about miles being a measure of long distances. Suppose someone tells them the library is 3 miles down the road. Will they be able to judge whether they can walk that distance? Many of us judge longer distances by how long it takes us to get somewhere—driving or walking. Students can use time as a cue also. Of course, we can also acquaint them with kilometers being a similar measure used for longer distances in the metric system.

WALK A MILE

OBJECTIVE: The student will be able to relate a mile to the time it would take him to walk it. The student will also relate a long distance (up to 100 miles) to the time it would take to drive it (or ride in a car or bus for that time).

MATERIALS:
- A watch
- Knowledge of trips the students have taken such as to a grandparent's house (call parents ahead of time)

SUCCESS STEP: Ask the students if they can walk a mile.

PROCEDURE:
1. Find a house or store or park area that is about a mile from your house or school. You can probably measure it best in your car. If your students have difficulty walking long distances, choose a destination that is a half mile away so that the walk there and back will be one mile.

2. Take the students for a mile (or two-mile) walk. Note the time that it takes to walk one mile.
3. Talk about the walk with the students. Tell them that they walked a mile (or two miles). Were they tired? How long did it take? Would they walk that far to buy a snack? Talk about other places that are one mile away. Find more places that they can walk to and compare the distances with this one-mile walk. Relate the time that each walk takes.
4. Talk about riding in a car. How long do they think it would take them to go a mile in a car? Compare the time it takes to walk or to ride in a car for one mile.
5. Discuss a trip you have taken—how many miles and how long it took. Have the students describe a trip they have taken. With your "inside knowledge," you can relate the number of miles and the travel time for the students' trips.
6. The concept of miles, distance, and time will only become useful for the student if it is generalized to many situations. Parents should routinely talk about the miles traveled when they take family trips and relate that to the time needed for traveling. Teachers can plan short walks or field trips and then talk about the miles traveled. When discussing news events, parents or teachers can talk about how many miles away the event occurred.
7. Use MapQuest or another site on the Internet to calculate the distances between a student's home and places she would like to go (California, Six Flags, etc.). Talk about how far that is and whether she could walk there, could drive there in a car, or would need to take a plane, etc. Many students with Down syndrome and other hands-on learners do not have a good concept of distances. I think that it is not a really a math problem, but perhaps more a problem related to insufficient experience. You can start with a map showing where the student lives, expand to the state where she lives, and progress to neighboring states and then to the country itself, but that is another whole subject.
8. Discuss miles as compared to the metric system of kilometers. 1 mile equals about 1½ kilometers. Tell students that kilometers are smaller than miles, but measure long distances like miles do. Try walking one kilometer and then continuing on until you have walked a mile. How much longer did it take to walk a mile rather than a kilometer?

GENERALIZATION ACTIVITIES
- Follow the student as she rides her bike or takes a walk in the neighborhood and figure out approximate miles to various places. Have the student tell how many miles she wants to go when she asks permission to ride her bike.

- View or enter a run such as the 3 mile run or the 5K (kilometer) run.
- Use a pedometer to measure walks or to measure how far a person walks doing everyday tasks.
- When visiting a park that has measured trails marked, talk about whether you want to go (or have time to go) on the 1 mile trail to the lake vs. the 2 mile trail to the woods, etc.

Perimeter and Area

Think back to the last time you needed to know the area or perimeter of something in your life. Maybe you were comparing the square footage in one house you were considering buying with the square footage in another and asked the real estate agent for the figures. Or perhaps you were buying new carpet or tile for your floors, or you were having a fence installed. Perhaps you went ahead and figured out how many square feet of carpet you needed yourself, but then again, perhaps the flooring company sent a representative out to your house to determine how much carpet you needed.

Clearly, adults do need to figure out area or perimeter occasionally, but just as clearly, they can go months or even years without needing to do so. And in most cases, it is perfectly acceptable to ask the professional (carpet or fence salesman, etc.) to figure out the area or perimeter for you. Still, it is important to understand the difference between area and perimeter so we can be knowledgeable consumers of products. Most people with Down syndrome and other hands-on learners do not have to figure out perimeters or areas for themselves. However, they may need to understand the difference between perimeter and area to understand others' explanations.

INTRODUCING PERIMETER

OBJECTIVE: The student will demonstrate the meaning of "perimeter" by explaining how to measure it.

MATERIALS:
- Tape measure or ruler
- Rectangular objects to measure
- String or ribbon
- Slate or whiteboard
- Perimeter Problem Worksheet (Appendix B)

SUCCESS STEP: Ask the student if walking around the room once would be a mile. Praise correct answers or tease her about walking a whole mile in the room. Tell her that we measure lengths and widths of a room differently from distances in miles.

PROCEDURE:
1. Have two students (or adult and student) use a tape measure to measure a desk or table. One student holds the case while the other handles the tape. Write

the measurements on the slate or whiteboard, such as 20 inches wide by 36 inches long.

2. Ask the students if you will have enough string to go around the desk if you cut it so it has the measurements on the whiteboard. Take a string and measure off the length and then the width (for example, 36" and then 20").

3. Bring the string to the table or desk and show the students that the string only goes halfway around.

4. See if you can get the students to say that they need another string the same length. Then cut another length of string and have them help you put the string all around the table. Tell them that there is a name for the "distance around" called *perimeter*.

5. Have students measure the perimeter of books or other rectangles in the room by wrapping a string around the object and then measuring the string with rulers, yardstick, or tape measures.

6. Talk to the students about how to use perimeter in real life by explaining how to use perimeter in the following word problem:

> *Mr. Gabriel has a little dog named Andy who keeps running away. Mr. Gabriel has decided to put a fence around part of his yard and make a play area for Andy. He is going to draw a picture of the fenced-in play area. He wants the play area to be 6 feet long and 5 feet wide. So he draws a rectangle that is 6 inches long and 5 inches wide. (1 inch in the drawing stands for 1 foot in his yard.)*

7. Draw a picture of a rectangle on the board and label the length 6 and the width 5. Ask a student to label the two unlabeled sides. Ask the students how they can figure the amount of fence Mr. Gabriel needs. They will probably add all four sides together. Some students may figure out that they can multiply the length by 2 and the width by 2 and add them. Any accurate way is fine. Eventually, you should come up with an answer of 22 (inches/feet). Tell them they have figured out the perimeter.

8. Tell the students that when they hear the word perimeter, they should think fence.

9. Have the students complete the Perimeter Worksheet from Appendix B.

INTRODUCING AREA

OBJECTIVE: The student will be able to compare the perimeter and the area of a rectangle.

MATERIALS:
- The picture of Andy's Play Area from Appendix B
- A string about 24 inches long
- 30 cardboard or paper squares 1 inch square (green, if possible); model in Appendix B
- Perimeter Dond Area Worksheet (Appendix B)

SUCCESS STEP: Ask the students to run their finger along the perimeter of the fenced-in area in the picture. If they are successful, praise them; if not; model for them and have them try again.

PROCEDURE:
1. Tell the students a continuation of the story of Mr. Gabriel, above.

   Mr. Gabriel now has a fence to keep his dog from running away. But when Andy stayed in his play area, he pulled up most of the grass that was planted there. Mr. Gabriel decided to have new sod (long rolls of dirt where grass is already growing) planted in the play area. How much grass sod should Mr. Gabriel buy? Is that the same amount as the perimeter?

2. Remind the students what they figured was the perimeter of the play area (22 inches standing for 22 feet).
3. Talk about the sod needed to cover the area. A string can't measure the whole area. Use the paper squares to cover the rectangle. Have the students count the number of squares they use (30 square inches). Write the amount of squares as the area on the whiteboard.
4. Hold up the string and contrast it to one of the squares. Tell the students that the string is long and measures in inches. However, the area is square, not long, so it must be measured in square inches.
5. Write on the slate or whiteboard:

   Perimeter is around the **outside** of the rectangle (fence)
   Area is all the **inside** of the rectangle (grass sod)
   Perimeter is measured in **inches or feet**
   Area is measured in **SQUARE inches or SQUARE feet.**
   To find perimeter you usually ADD
   (Length + Length + Width +Width = Perimeter
   To find area you MULTIPLY
   (Length × Width = Area (square)

6. Talk over the above principles with the students, but it is not important if they understand how to figure perimeter and area at this time.
7. Have the students do the Perimeter and Area Worksheet from Appendix B. Work each problem together. For each problem, have the students use a marker or colored pencil to trace around the perimeter with one color (say, red) and color in the area with another color (say, blue). Then, if appropriate, have the students calculate the perimeters and areas of the rectangles. The major purpose here, however, is to acquaint them with the vocabulary of *perimeter* and *area* and give them a visual picture of what the difference is—not necessarily for them to memorize how to compute perimeter and area.

GENERALIZATION ACTIVITIES: Possible suggestions for generalization activities:
- Measure heights of family members or class members.
- Measure photos or posters to see what size frame you need to buy or whether they will fit in a scrapbook.
- Help measure windows for curtains.

- Parents can have their child measure furniture in her own bedroom before rearranging it. For example, will the bed fit by the window?
- If you are shopping for new furniture, let your child help you figure out how many square feet you have for a new desk, computer, etc.
- If you are shopping for a rug or carpet, let your child help you measure the area that needs to be covered.
- Measure the waist and the inseam on a pair of jeans that fit well. Tell your child what her size is. When shopping for new jeans, bring a small tape measure and measure the waist and inseam of the jeans that you try on or look at the inseam lengths marked on jeans. If there are no jeans that are exactly the right length, point out jeans that are too long or too short and talk about how the jeans would need to be shortened if you bought them.
- Relate sports to measurement by having the student "walk off" the distance between the yard lines at a football field or compare the distances necessary for a baseball player to hit an over-the-fence home run in various ball fields.
- When shopping, point out items that are measured in square feet or square inches (wrapping paper, plastic wrap, tape). Do you get more if you buy one brand vs. another?
- Measure and cut Contact paper or shelf paper for drawers or shelves.
- Measure and cut newspaper to go on the bottom of a hamster or bird cage.
- Before a haircut, help the student figure out how much shorter she wants her hair—a half an inch? two inches?

Capacity (Volume)

If the students have finished the chapter on fractions, they will have experience with measuring by ¼, ½, ¾, ⅓, ⅔, and 1 whole cup. They will also be able to use ¼, ½, ¾, and 1 whole teaspoon and 1 tablespoon. First they learned by measuring rice and then by using baking ingredients in an actual cooking situation.

You may want to review the cups and teaspoons, emphasizing the equivalents this time, such as two ¼ cups = ½ cup, two ½ cups = 1 cup, etc. I think that it is easier to explore the equivalents by pouring rice than with water, as is usually done. One equivalent was not taught in the fraction chapter: 3 teaspoons = 1 tablespoon and you may want your students to experience that fact now.

In exploring capacity further, we need to look at larger measurements commonly used in the U.S., including pint, half gallon, and gallon. In addition, a more long-term project on volume may be to make a rain gauge.

CUPS, PINTS, QUARTS, AND GALLONS

OBJECTIVE: The student will be able to identify cups, pints, quarts, half gallons, and gallons and be able to demonstrate their equivalents.

MATERIALS:
- Real examples of a pint (whipping cream)

- Quart (orange juice or milk)
- Half gallon (milk, lemonade)
- Gallon (milk)
- Juice drink that is one cup in capacity
- Measuring cup
- Empty cartons
    - cup
    - pint
    - quart
    - half-gallon
    - gallon
    (Cut off the tops of the bottles or cartons so you can easily pour rice into them)
- Rice, dried beans, or sand
- Slate or whiteboard and marker or chalk

SUCCESS STEP: Ask if anyone has heard the word volume before. (They have probably heard "turn down that volume!") This is a different kind of volume. This type of volume is the space inside a container. Show them the empty cup and say that the volume is the space inside the cup. Show them the empty quart bottle or carton and ask what is the volume inside this quart. Hopefully they will say it is the space inside the carton. (If not, tell them.) Have one student point to the volume of the quart by pointing a finger inside the quart container.

PROCEDURE:
1. Line up the real examples of capacity measurements in front of the students in size order.
2. Have the students guess how many measuring cups make a pint.
3. Using the empty cup and rice (or dried beans or sand), have the students put two cups of rice into the empty pint measure.
4. Write the phrase **2 cups equal 1 pint** on the board.
5. Then have the students guess how many pints make up a quart.
6. Using rice, have the students pour two pints into the empty quart measure.
7. Write the phrase **2 pints equal 1 quart** on the board.
8. Repeat the above procedure with the half gallons and the gallon. Write the phrases:
      2 quarts equal 1 half gallon   **2 half gallons equal 1 gallon**
9. Ask the students if they can figure out how many quarts there are in a gallon. Have a student demonstrate with rice, if necessary. Point out to students that "quart" is related to "quarter" or ¼. There are 4 quarts in a gallon just like there are 4 quarters in a dollar.
10. Be sure that the students do the measuring themselves. A demonstration by you is just not the most effective way for them to learn. If you have a small group, try to give every student a chance to do the measuring equivalents.
11. You may want to make an individual flap chart to help the students memorize these equivalents. (See below.)
12. You could also teach the students a mnemonic to remember the measures in capacity order: **c**up, **p**int, **q**uart, **h**alf-gallon, **g**allon. **C**an **P**riscilla **q**uit **h**er **g**iggling?

Flap charts can be used in many situations when students need to memorize facts. See page 98 in Chapter 8 for general instructions. Here is what to write on the flaps and on the paper under the flaps:

Front Flap	On Back Paper
1 pint = _____ cups	2 cups
1 quart = _____ pints	2 pints
1 quart = _____ cups	4 cups
1 gallon = _____ quarts	4 quarts
1 half-gallon = _____ quarts	2 quarts

The student studies by looking at the phrase, lifting up the flap, and reading the answer. She then puts the flap down so the number doesn't show and repeats the answer. The student then practices all the equivalents until she can say them without assistance.

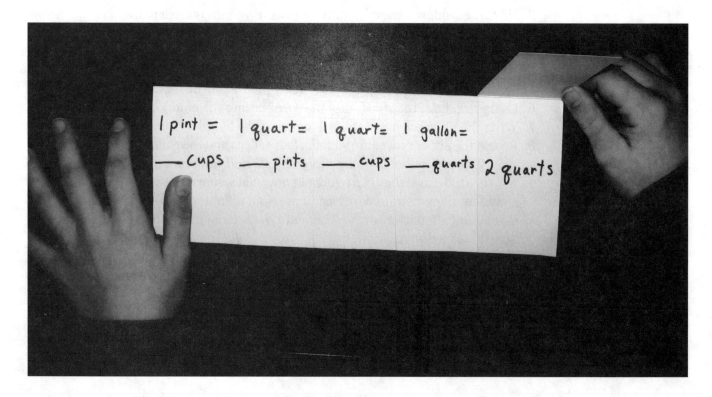

MAKING A POWDERED DRINK (PUNCH)

The best situation for this activity would be an occasion where the students would really be responsible for making drinks for a group of people such as a picnic or a party. If you don't have a real situation, you may have to make a situation. Perhaps you could have a lemonade stand or sell punch to the school for a pep rally; make a drink for a church or another organization; or have a class party. The idea is to show how learning math can be useful to them. You do not have to use Kool-Aid or a similar powdered drink; you can make a juice punch or any drink that needs measurement of cups, quarts, and gallons.

OBJECTIVE: The student will be able use knowledge of liquid and solid measures in a useful, real world situation.

INGREDIENTS:
- 1 package Kool-Aid or other powdered drink mix
- 1 pint (or two cups) 7-Up or similar beverage
- 1½ to 2 cups of sugar
- 2 quarts cold water
- 1 pint (or two cups) orange juice

MATERIALS:
- Punch ingredients as described above.
- Measuring cup
- Quart jar
- Pitcher (more than 2 quart size)
- Half gallon container
- A gallon bottle or large punch bowl, if large amount of punch is needed

PROCEDURE:
1. Pair up students to make the punch if you have more than one student to teach.
2. First have a student measure the sugar and put it in the pitcher.
3. Have another student put the powdered drink mix with the sugar in the pitcher.
4. Then have a student add one quart of cold water.
5. Say, "We put one quart of water in the bottle. We need one more quart of water. Do you know how many cups would be in that quart?" Have the students measure cups of water into the quart jar and count them. Then have them put the rest of the water in the pitcher.
6. Have the students guess whether all the punch will fit in an empty half gallon. The adult or older student should then pour the drink from the pitcher into the half gallon carton to show that two quarts equal one half gallon.
7. Then the adult or older student should pour the drink back into the pitcher or punch bowl (a pitcher or bowl that will hold more than 2 quarts).
8. Have one student measure out 2 cups or 1 pint of orange juice and pour it in the pitcher and stir.
9. Have students measure out 1 to 2 cups of 7-Up or another lemon-lime drink into the punch. Add ice and serve to students or others.

This activity can be used for making larger quantities of punch. Several groups of students can make the recipe individually and then combine it. You can even use multiplication or repeated addition to make the large quantity all at once.

Temperature

Teaching Math, Book 1 introduced some basic information about temperature with the goal that students would be aware of the concept and have a general understanding of what higher or lower temperatures meant. In this book, we will cover some of the everyday uses of temperature measurement that adults need to understand in order to live safely in their environment.

TEMPERATURE AND THE WEATHER

One of the best ways to teach the practical uses of measuring the temperature is to have an indoor-outdoor thermometer that can be seen from a window. Each day the family or the teacher can comment on the temperature and the weather. It also helps to identify some points that have meaning for the student. For example, you can say that at a certain temperature, everyone needs a coat and that at another temperature they need a sweater, or that it has to be a certain temperature to go swimming.

OUTDOOR TEMPERATURES

OBJECTIVE: The student will relate outdoor temperature to daily living activities such as deciding what clothing or outdoor activities are appropriate at certain temperatures.

Note: The Celsius temperature scale is mentioned for those outside of the US. I am not recommending that students learn how to convert from Fahrenheit to Celsius.

MATERIALS:
- Chart of Fahrenheit and Celsius temperatures and related activities (below)
- Optional—meat thermometer

PROCEDURE:
1. Discuss the temperature chart with the students. Talk about clothes that are appropriate at different temperatures. Discuss appropriate activities.
2. (Optional).You might want to use a meat thermometer to check the temperature of ice and boiling water. (See food safety section below.)

Fahrenheit	Description	Celsius
32 degrees	Water freezes	0 degrees
68 degrees	Comfortable indoor temperature	20 degrees
90 degrees	Hot weather outside	32 degrees
98.6 degrees	Normal body temperature	37 degrees
102 degrees	Fever in body	39 degrees
212 degrees	Water boils (sea level)	100 degrees

FOOD SAFETY

The temperature of the food we eat also affects daily life. Bacteria or germs are always in our food. However, they grow very fast when the temperature is between 45 degrees F. and 140 degrees F. (7—60 degrees Celsius). That range of temperatures is called the Danger Zone. Cooked or refrigerated foods left at Danger Zone temperatures for over two hours can make people sick. Most foods should be kept at 39 degrees F. (4 degrees C.) in your refrigerator. Meat and poultry should be kept colder, at about 34 degrees F. (1 degree C.). Even when kept at the proper temperature, however, foods kept in the refrigerator will gradually spoil.

Food stored in a freezer at 0 degrees F. (-18 degrees Celsius) will be safe. Freezing keeps the bacteria or germs from growing. Cooked meats and other foods should be cooked until their temperature is above 140 degrees F., to a temperature of about 150-165 degrees F. (65.5 to 74 degrees C.).

IS IT SAFE TO EAT?

OBJECTIVE: The student will be able to measure the temperatures of foods to determine their safety.

MATERIALS:
- A meat thermometer (available at stores for about $6-7)
- Food Temperature Danger Zone chart (Appendix B)
- Packages of frozen food, ice cream, hot dog (cooked), and foods in a refrigerator (getting these materials will certainly be easier for parents at home, but some teachers may be able to get access to the kitchen of the school cafeteria)
- Optional: hamburger patty cooked and ready to eat

SUCCESS STEP: Have the student point to several temperatures on the meat thermometer and read them out loud.

PROCEDURE:
1. Discuss the facts about bacteria (germs) growing in food that could make you sick. Explain the Danger Zone using chart from Appendix B.
2. Have the students measure the temperatures of various foods by inserting the pointed end of the meat thermometer in the packages. Compare them with temperatures in the Danger Zone:
   - Ice cream (or other frozen food) right out of the freezer (should be close to 0 degrees Fahrenheit/minus 18 Celsius)
   - Ice cream(or other originally frozen food) when allowed to sit out at room temperature for an hour
   - Most food stored in refrigerator (should be 40 degrees F./4 C. or colder)
   - Meat stored in refrigerator (should be at about 34 degrees F./1 C.)
   - Hot dog cooked in microwave or boiled (should be 150 degrees F./65.5 C.)
   - Optional—Cooked hamburger (should be 155 degrees F./68 C.); ground meat requires a higher temperature than a hot dog.
   - Let some food sit at room temperature and take its temperature every half hour. Make sure that the students understand that 2 hours is the absolute limit for perishable food to not be refrigerated.

ILLNESS

Another time that children hear about temperature is in connection with illness. Secondary students should be able to understand about raised temperature by learning about the body and its reaction to illness in a health unit. Some students will be able to read a fever thermometer with practice. You can have the students take their own temperatures with a fever thermometer. Be sure to have alcohol available for cleaning the thermometer if more than one student needs to use it.

GENERALIZATION ACTIVITIES
- When unpacking groceries from the store, show your child how you know whether something needs to go in the refrigerator or freezer. Then have your child help put them away accordingly.
- Before you leave the house for school, to go shopping, etc., discuss the predicted highs and lows for the day and how that affects what you wear. The weatherman says it's going to be 50 degrees for a high. Should you bring a jacket?
- At home, have the student take her own temperature every morning at the same time and record the temperature. You figure out what her average body temperature is, and then let her know what temperature means that she has a fever. For older students, talk about distinctions between: (a) a fever that means she should stay home and rest in bed; b) a fever that means she needs to see a doctor.
- Teach your child how to set the oven temperature at home (as well as proper precautions to take when putting in/taking out food).
- At home have your child help read a meat or candy thermometer if you use one in cooking. When you are using a meat thermometer, show the student how it works. If you have a digital one where you set the ending temperature and it also shows the current temperature, you can use "how much more" subtraction to figure out how many more degrees the meat needs to be.
- At school or home, have the students choose a city in a different state or country and track the high and low temperatures for a week or more. (Look in the newspaper or online for weather information.) How much warmer is it in Honolulu in December than in your city? How much colder is it in Toronto in May? When you are having winter temperatures in your area, what is the temperature like in Sydney, Australia?

Comparisons with the Metric System

LITERS AND GALLONS

OBJECTIVE: The student will compare common units of capacity in the metric system with those used in the USA.

MATERIALS:
- An empty half-gallon container with the top cut off so water or rice can be easily put into the container
- An empty 2-liter container (soda bottle) with the top cut off
- Quart container and 1-liter container (optional)
- Water or rice to measure

PROCEDURE:

1. Show the students an empty half-gallon bottle and the empty 2-liter bottle.
2. Ask the students if they think these two bottles could contain the same amount of water or rice. Tell them that the soda bottle is one of those mysterious ways to measure things used by scientists. This bottle contains 2 liters. *Liter* is the name given to this amount of water or rice in the metric measurement system.
3. Have the students repeat the word *liter* several times and then find the word *liter* on the bottle label.
4. Have the students pour water or rice from the two-liter bottle into the half-gallon container so they can see that the two measures are very close to each other.
5. Optional—Use an empty liter bottle with a quart container and compare the volume of a quart and liter by pouring rice and water from one container to the other.

OUNCES, GRAMS, AND POUNDS

OBJECTIVE: The student will be able to recognize ounces, grams, and pounds as measures of weight.

MATERIALS:

- A postage, food, or diet scale that has both type of units on it, if possible (or two separate scales, if you have them)
- Small items to weigh
- Cans or packages of food with both ounces and grams listed
- Small bathroom scale

SUCCESS STEP: Ask the student what the scale is for. Praise correct answer or demonstrate weighing.

PROCEDURE:

1. If you have access to a postage or food scale with ounces and grams, you should weigh the same item first in ounces and then in grams. Help the students to understand that the item doesn't weigh more when it is weighed in grams vs. ounces. You are just using a different unit to measure its weight.
2. If you don't have access to a metric scale, look at some grocery items that have weights in both grams and ounces and compare them. Which is bigger? An ounce or a gram?
3. Optional—If it is important in your area for the students to know both methods of measurement, you may make a flap book as described on page 176 with familiar units on the top flap and metric units on the bottom area.
4. First ask the students if they can weigh themselves in ounces. If they answer incorrectly, tell them that they need a larger measure called pounds. Using the bathroom scale let the students weigh themselves. Position the scale so no one but the student can see her weight. Emphasize that we weigh small things with ounces and grams. Larger objects such as people are weighed in pounds.

Time

To live independently or semi-independently as adults, students generally need only to be able to tell time as so many minutes after the hour. This is how time is most often expressed in writing—for instance, on bus schedules, class schedules, or in the TV guide—and also how appointment times (for doctor's visits, job interviews, etc.) are usually expressed. Your child or student does not need to tell time as so many minutes before the hour sometimes and after the hour at other times. She needs consistency. However, even though we want students to be consistent in *telling* time, they may have to learn to *understand* the other ways we tell time, including:

This student made a personalized cuckoo clock with a photo of himself as the cuckoo

- Expressing time as minutes before and minutes after
- Telling time by the quarters
- Using different vocabulary such as *ago, half past, almost, in __ minutes*

MINUTES BEFORE AND AFTER

OBJECTIVE: The student will be able to demonstrate understanding of minutes *before* and *minutes after*.

MATERIALS:
- A teaching clock, either commercially made or made from a paper plate or piece of cardboard with the hour and minute hands fastened with a brass fastener.
- Post-it notes (two colors)

SUCCESS STEP: Using the teaching clock, have the student tell the time for 6:05, 6:10, and 6:15 when the teacher sets those times. If she is not successful after a short review, refer to Chapter 17 in Teaching Math, Volume 1, for ideas on teaching her to tell time.

PROCEDURE:
1. Tell the students that the way they have told the time is correct. However, other people have different ways of telling the same time. They talk about *minutes before* and *after*.
2. Put a Post-it note that says "5 minutes after 6" on the teaching clock near the 5 minute mark and put the hands on the clock at 6:05. Have the students repeat the time: "5 minutes after 6."
3. Repeat the above procedure for 6:10 and 6:15.
4. Continue having the students match the standard times with the *minutes after,* up to 6:30.
5. Repeat the procedure with other hours.
6. See if the students can set the teaching clock to certain times *after* the hour as told by the teacher.

MINUTES BEFORE

When you think that the students are secure in understanding the *minutes after* way of telling time, you may introduce the *minutes before* time telling. Telling time *before* the hour is much more difficult because the numbers are not the same as when telling

time the standard way. The student has to skip count 5's in the opposite direction of what he has been taught. It is possible that the student will have a difficult time understanding the *minutes before* way of telling time. However, this is worth working on because being able to understand minutes (or hours) *before* important events helps you understand how much time you have to get ready for something. For example, if you only understand the time as 8:40 and not also as 20 until 9, it may be harder to figure out how much time you have to get ready if you need to leave at 9:00.

I am convinced that students need to be able to tell time with an analog clock, as well as with a digital clock. With an analog clock, a student can see visually that the minute hand is getting closer to the next hour. That visual cue is important for hands-on learners. I also recommend using a device called a Time Timer (www.timetimer.com). When you set this timer, a red area on the timer gets gradually smaller as the allotted time passes. It can be used to help students understand how much time they have to do their homework, to get ready in an appropriate amount of time, and other situations where it is useful to see time passing in small increments.

MATERIALS:
- Teaching clock (as above)
- Post-It Notes.

SUCCESS STEP: Have the student count by 5's around the clock, going clockwise.

PROCEDURE:
1. Tell your student(s) that some people like to tell the time before the hour. After 6:30, they count the time to the next hour. Show them 6:55, and ask what o'clock it's going to be soon. If they can say it's going to be 7:00, say "Yes, that's right. It will be 7:00 in 5 minutes." If they don't know that it's going to be 7:00, show them on the clock that once the minute hand moves 1, 2, 3, 4, 5 minutes, it will be 7:00. So, right now, it is 5 minutes BEFORE 7:00.
2. To check that the students understand, put the clock on a variety of times when it's 5 minutes before and see if they can say the time. Then set the clock to 10 before the hour, etc
3. Using Post-It Notes, write 25 minutes before 6, 20 minutes before 6, and 10 minutes before 6 on separate Post-It notes. See if the students can match the Post-It notes

with the time on the teaching clock. We are just trying to get them to recognize the *minutes* way of telling time, not to have them be able to tell time with the *minutes* before *and after* method.

4. Repeat the above procedure for different hours.

TELLING TIME BY THE QUARTER AND HALF HOUR

OBJECTIVE: The student will be able to recognize quarter past, quarter 'til, and half past.

MATERIALS:
- Teaching clock as above
- Quarter Clock (Appendix B)

SUCCESS STEP: Ask the student, "Do you remember another way to say one-fourth? That's right, one quarter." If she is not successful, give her a quarter and ask what it is. Then ask, "Do you know how many quarters are in a dollar?" If she doesn't know, tell her four quarters in a dollar, so one quarter is the same as one-fourth of a dollar.

PROCEDURE:
1. Using the teaching clock, have a student show you how she would make the clock hands go the *whole* way from 12 o'clock to 1 o'clock. Have her recognize that the minute hand needs to go all the way around from 12 back to 12.
2. Once the student can make the hands go the whole way around, ask her if she can make the minute hand go only half way around. Show her the Quarter Clock and point to the 6 and the half past label. Then show her how when it is 12:30, or half past 12, the hands of the clock are straight up and down, dividing the clock in halves. Whenever the minute hand is on the 6, it has gone halfway around the clock with 30 minutes on each side.
3. Ask the student how many parts the circle has been divided into on the Quarter Clock (4).
4. Have the student point out where the minute hand would be if the time was a quarter after. Repeat with quarter 'til.
5. Practice these common time phrases during activities in the day.

MORE TIME VOCABULARY

OBJECTIVE: The student will understand other commonly used time vocabulary words, such as almost, ago, and in ___ minutes.

PROCEDURE:
1. Try to teach the idioms of time using a charade-type game. Some suggestions are:
   - Run around acting as if you are late. Say, "I'm going to be late. It is *almost* 8:00." Ask the student to put the hands of the teaching clock at a time that would be almost 8:00.

- Act unconcerned about a problem that someone is telling you about and say, "That happened such a long time *ago*. I don't really remember what happened." Ask the student whether ago means tomorrow or last night. Then ask if *ago* means that it happened that morning or the week before.
- Act as if you are angry because you have waited too long for someone else. Say, "Oh, he is so late. He told me *half past* 8:00 and I have been waiting here forever. It is 8:00 now. Why is he so late?" See if the students can tell you what the error is. Have one of them make the teaching clock say half past eight.
- Act frustrated and say, "In a minute, in a minute—that means one minute, right? Well, I have waited 9 minutes already and she is still talking." See if the student can tell you what "in a minute" really means in our society—or in our families.
- Set the clock at 6:15. Say, "*Quarter past* 6. It is *quarter past* 6. I guess I have to give you a quarter." See if you can get the student to tell you what is wrong and correct you on the meaning of *quarter past*. Repeat for *quarter before*.

2. Listen to the students and see if there are other ways that we refer to time that they do not understand. Act out the situations and try to make them exaggerated so they are funny. That will help the students to remember the vocabulary.

My son Scott tells time the way it is expressed on a digital clock. I asked him what he does if he doesn't understand when someone tells him that it is so many minutes before the hour. He said, "I just ask them to tell me the right way." I have never seen him have trouble misunderstanding a time told to him, so his way must work. If a student is not able to understand "minutes before," he still may be able to function well as far as time goes.

AWARENESS OF TIME NEEDED FOR ACTIVITIES

Sometimes a student can tell time on the clock but does not really have an internal feeling for how much time is needed for an activity. For example:

Mom says, "You have 5 minutes before the bus comes."
McKenzie says, "Is 5 minutes a long time or a short time?"

I would first try a Time Timer for some daily activities so the student can see the red area get smaller as time passes. See the explanation about the Time Timer on page 183.

Also try the matching game below to help the student understand how long various activities tend to last. You may disagree with the times I have listed. Change the times so they are appropriate for your student(s).

TIME MATCHING GAME

OBJECTIVE: The student will be able to match appropriate times with daily living activities.

MATERIALS:
- Activity strips cut from Time Matching Game (Appendix B)

- Time cards cut from Time Matching Game
- Answer sheet for Time Matching Game

SUCCESS STEP: Ask the student to put the activity strips all together in one pile.

PROCEDURE:
1. Put the time cards in 3 rows of 4 each. At first put them face up. Later, if you want to make a memory game, you can place them face down.
2. One student picks up an activity strip and reads it out loud. She then picks one of the time cards that are spread out in front of her. If she is correct, she puts the activity and the time card together and keeps it near her. If she is not correct, she puts the activity card at the bottom of the pile and the time card back where it was originally.
3. The winner is the person with the most pairs at the end of the game.
4. You can add activities and times to fit your own family or student.

GENERALIZATION ACTIVITIES

Nathan got a ride to school from his dad. He was usually not ready when Dad needed to leave. Everyone in the family tried to get him ready on time, and there was often shouting and angry words in the morning. Nathan could tell time, but he didn't take any responsibility for monitoring his time as he got ready in the morning.

Eventually his family decided that he had to be responsible for getting ready on time. A short schedule was posted with times for showering, eating breakfast, and getting outdoor clothes on. Nathan was told that he alone was responsible for getting ready in the morning. Of course, the first day he wasn't any faster. He also had to walk to school and was late. In a week, he finally started looking at his watch as he got ready. Gradually he took more responsibility for being on time—and early morning at Nathan's house was much quieter.

There are many opportunities to use knowledge of time in everyday living. Most students will be able to use a wristwatch to follow a daily schedule. Analog watches should have large, easy-to-read numbers. Digital watches should not have many other figures in the display space along with the hour, minute, and seconds.

- Give the student responsibility for knowing when to turn the TV on for her favorite shows.
- When the student has to wait for some time, don't say "in a minute" but tell her a specific time (e.g., "We'll leave at 5:15"). Then be sure to be ready when she tells you that it is that time.
- Use specific times on a schedule at home or school and make the student responsible for being at the proper location on time.
- Time how long it takes her to shower, dress, and get ready to go somewhere. Then discuss with her how much time she needs before she is due to be somewhere. Praise and possibly reward her when she is ready on time for an event.

- If she is learning about bus transportation, show her the bus schedule and explain how it works. Gradually have her take responsibility for getting to the bus stop on time.
- Get your child her own alarm clock and teach her how to set the time and the alarm.
- Teach her how to set the clock on the VCR, the microwave, etc. For practice going from analog time to digital time, have her refer to an analog clock when setting the time on the microwave, etc.
- Teach her how to set the time on the micro-wave—you can use 60 (seconds) or 1:00 for a minute; 2 minutes is 2:00, not just 2—that is only 2 seconds, etc.
- Point out when other people use the idiomatic ways to tell time. Make sure that the students understand what the standard time is when time is expressed in an idiomatic way.

Games to Buy or Try

- **Math around the Home** (International Playthings; available from educationallearninggames.com)— Players progress around a game board answering practical questions related to time, temperature, money, and measurement.
- **Conceptual Bingo—What Time Is It?** (Conceptual Math Media). A bingo game in which the caller can choose from 6 different levels of difficulty (for example, the 6:45 card may be read as "quarter 'til 7," "3 hours before 9:45," etc.).
- **Timing It Right** (Learning Resources). Players move their pieces around a game board and land on squares printed with a digital time, then must set an analog clock to match.

CHAPTER 13

General Concepts of Money

Questions to be answered:

Can the student:

1. Count currency by multiplying and then adding to get the total.

2. Skip count as used in counting currency.

3. Round off prices to the next highest dollar.

4. Buy one item using the next-highest-dollar strategy.

5. Read prices correctly using dollars and cents.

6. Name the commonly used coins and their values.

7. Skip count coins.

8. Use pennies accurately.

9. Figure the total amount of money for buying several items, using the next-highest-dollar strategy.

Teaching Math, Book 1, Chapter 20 discussed money in quite a bit of detail. The student working in *Book 2* should already know the concepts of paying and receiving, be able to identify currency (bills) from $1 to $50, and identify the value of commonly used coins. He should also have some experience in using the next highest dollar system for buying items.

In this chapter, we will focus on practical money skills needed for such activities as counting coins and currency and add more information on paying for items using the next highest dollar strategy (rather than counting out exact change).

Jake wanted to buy a DVD that was priced at $14.99. He paid with a twenty-dollar bill and got change back. When his mother asked him how much the DVD was, he said, "Twenty dollars!" Because he paid with a twenty-dollar bill, he considered that the cost was twenty dollars. He didn't figure in the fact that he received change back. His mother had him role-play the situation again. She had him count the change. Then they used the calculator to subtract the amount of change from the twenty dollars he used to pay. His mother emphasized what the real cost of the DVD was. She repeated this teaching each time he paid for an item.

Counting Currency

Some of the activities in this book require the student to figure out how much money (currency) he has left. There are two different ways to count currency that are useful in different situations—skip counting and a more formal method of using a calculator to multiply and add. To use the calculator method, you count up the number of each type of bill and multiply by the value of that bill. "I have 7 five-dollar bills, so I will multiply 7 × 5, which equals $35. I have 2 ten-dollar bills, so I will multiply 2 × 10, which equals 20 dollars. Then I need to add the two totals together to get my final amount…. I have $35 dollars in fives and $20 in tens, so I have $55 altogether." When there is a lot of currency in different denominations, it is usually best to use this calculator method. You can then check your calculations if needed.

The Earn and Pay game used in the informal assessment in both Book 1 and Book 2 includes a money total slip that requires the student to multiply the number of bills of each denomination and then add his results to get the total amount of money. Review those concepts by playing the Earn and Pay game frequently.

EARN AND PAY GAME

OBJECTIVE: The student will be able to total his money at the end of the game using the multiplication and then addition strategies as done on the Money Total slip.

MATERIALS:
- The Earn and Pay game from Appendix B
- Earn and Pay cards copied on cardstock from Appendix B
- Money (play) $1, $5, $10, and $20 bills—enough for $85 for each person. (Smaller versions of currency, photocopied only on one side, are available in Appendix B, although realistic play money is preferred)
- Dice
- Game markers (small buttons, tokens, coins, etc.)

PROCEDURE:
1. Copy the game board from Appendix B and paste the pages on the inside of a file folder.

2. Cut out the Pay and Earn cards. Photocopy the Pay cards on one color paper and the Earn cards on another color, or paste them onto different colored construction paper. Put the card face down on the game board as indicated.

3. Give each player $85.

4. The first person rolls the die and moves his marker that number of spaces. The space landed on will be either a Pay space or an Earn space. The player picks up the top card from either the Pay pile or the Earn pile and receives or pays the amount listed on the card.

5. The corner squares involve some kind of direction, such as lose a turn or go back 2 spaces. If a player moves up or back following one of those directions, he does *not* pay or earn what is on that square.

6. The first person to reach the finish line is Winner I. He must throw the correct number to land on the Finish Square.

7. Winner II is the person who has the most money left. The students total up their money by using the Money Total slip (Appendix B3). That requires them to count the number of bills in each denomination, multiply by the value of those bills, and then add those totals to find the final answer. Students should be able to use the Money Total slip to get the total amount of money for determining the second winner in the game. When this slip is first used, you should plan for more time to explain how to use the form. You should also check the results.

SAMPLE OF MONEY TOTAL SLIP

Name:

Name of bill	Number of bills	Total amount of money in this bill
$1.00	13	$13.00
$5.00	3	$15.00
$10.00	2	$20.00
$20.00	2	$40.00
$50.00		
	Final Total:	$88.00

Skip Counting Currency

When we are dealing with large or various denominations of money, the multiplication plus addition way of computing a total is probably the most accurate method. However, most people count the money in their wallets by skip counting to get the total. In order to skip count money, the student needs to know how to count by 5's, 10's, and 20's. (If the student is not secure on counting by 5's, 10's, and 20's to 100, refer back to the multiplication chapter where this is discussed.) The skip counting currency method is shown below.

When an individual wants to find out how much money he has in his wallet or purse, he usually skip counts the bills as he moves them aside, one by one. We usually don't take the trouble to count up the numbers of different bills, multiply by the proper value and them add together all the totals (as described above)— even if a calculator is handy. Being able to skip count currency is a useful skill for everyday living, but not an essential survival skill for being able to handle money. However, students who have mastered the skills from *Teaching Math, Book I* will probably be able to do skip counting currency if given plenty of meaningful practice. Do not teach to the frustration level if the student is able to count currency using the calculator.

SKIP COUNTING FIVE- AND TEN-DOLLAR BILLS

OBJECTIVE: The student will be able to skip count 10-dollar bills up to $100, five-dollar bills up to $100, and twenty-dollar bills up to $100.

MATERIALS:
- 10 realistic ten-dollar bills for each student.
- 20 realistic five-dollar bills for each student.
- 5 realistic twenty-dollar bills for each student
- A wallet or purse

SUCCESS STEP: Give the student 3 ten-dollar bills, counting them "10, 20, and 30" as you do so. Ask the student to count them back to you in the same way. If he does not succeed, repeat the step using 2 ten-dollar bills. Praise a successful attempt.

PROCEDURE:
1. Repeat the success step with one more ten-dollar bill.
2. Put 10 ten-dollar bills in a wallet or purse. Say, "How much money do I have in my purse?"
3. Count out the 10 ten-dollar bills from your purse or wallet, 10, 20, 30, etc.
4. Put the money back in the purse or wallet and hand it to the student saying, "Now, how much money do you have in this wallet?" Have the student count it out loud for you.
5. If he has difficulty with skip counting, use the number line to illustrate or sing the numbers to a simple song like "Are You Sleeping" (Frere Jacques).
   Ten 20, 30… 40, 50, 60…
   70, 80, 90… 70, 80, 90…

Then a hun-dred, Then a hun-dred,
We are done… We are done.

Have the student sing the tune many times.

6. Repeat the same procedure with the five-dollar bills. First have the students skip count the bills to $25, then $50, then $100. Sing the 5s to a different tune such as "Ten Little Indians," "Ring Around the Rosy" or another tune that the students already know.

7. Repeat the procedure with $20 bills. Point out that the first digit in the $20 skip count is just like counting by 2s, e.g. 2, 4, 6, 8, and 10. If a student can't remember how to count by 20s but can count by 10s, show him the method of counting 20s introduced in Book 1. (Teach the student to touch each of Andrew Jackson's eyes as he counts "10, 20.")

Emily could easily count five-dollar bills to $100 and ten-dollar bills to $100. However, when she had both five-dollar and ten-dollar bills in her wallet, she would get mixed up and count incorrectly. When she had 3 five-dollar bills and 2 ten-dollar bills, she would count "5, 10, 15…20, 25." She didn't seem to be able to stop counting by 5's and continue with the 10's. Her teacher had her skip count by mixed 5's and 10's on the number line until she understood why her answer was wrong. However, it was still difficult for her to skip count mixed currency, so her teacher taught her instead to count only by 5's (counting two fives on the ten-dollar bills). She became very successful with this strategy.

See the counting money section in *Teaching Math, Book 1* for more detail on skip counting by fives.

SKIP COUNTING CURRENCY WITH MIXED VALUES OF BILLS

OBJECTIVE: The student will be able to count five-, ten-, and twenty-dollar-bills in the same group of currency.

MATERIALS:
- Realistic-looking one-, five-, ten-, and twenty-dollar bills for each student or the play money in Appendix B
- Number lines up to 100. This number line should be like a hundreds chart cut in lines of 10 each with the lines placed end to end. See Appendix B.
- Game piece for each student

SUCCESS STEP: Have the student line up the number line to 50. You can give him one number line of 10 at a time and just have him line them up straight. Assist him, if necessary, and praise his accomplishments.

PROCEDURE:
1. Each student should assemble his number line from 1-50 (or more, if possible) in front of him. Tell the students that they are going to take a little trip with jumps and a game marker.
2. Call out a jump of 5 or 10. Say, "Jump ahead 5" or "Jump ahead 10." Have the student make jumps with his game marker. Every time the game piece stops, the student should call out the number that he lands on.
3. Continue until the student can jump ahead five or ten numbers accurately. Begin again and hand out some twenty-dollar bills. Now call out some jumps of 20 as well as 5 and 10. (If it is hard for the student to mentally add by 20s, point out that he can make 2 jumps of 10 instead.)
4. Hand out the 5 and 10-dollar bills to each student.
5. Put the students in pairs (or the adult has to be a partner). One player calls out either "Jump 5" or "Jump 10" for the other player and puts currency to match that number on the table. The other player moves the game marker ahead the number of jumps called and announces what number he has landed on. If he is correct in the number of moves and the number reached, he gets to take the money from the other player. Then the roles are reversed. For example: Player 1 calls, "Jump five" and puts a five-dollar bill on the table by the number line. Player 2 moves the game marker 5 steps forward on the number line and calls out the number that he has reached. If he is correct, he gets to take the money away from Player 1. Player 2 then calls out the amount of money and puts it on the table while Player 1 moves the game marker ahead and calls out the number he lands on. If he is correct, he takes the money.
6. At the end of a specified period of time, end the game and let the players count their money.
7. When the players are secure at counting mixed 5s, 10s, and 20s, introduce one-dollar bills into the game. Tell the students that each dollar makes the total one number more. Repeat the pair game adding several single dollars.
8. When counting money at the end of the game, show the students that it is often easier to count your currency if you organize your money by denomination, from largest to smallest, and then skip count. It's definitely easier if you leave the ones until the end.

READING PRICES WITH DOLLARS AND CENTS

OBJECTIVE: The student will correctly read prices including dollars and cents.

MATERIALS:
- Copies of Place Value Money Chart (Appendix B)
- Reading Prices Worksheet (Appendix B)
- A teacher-made worksheet with prices under $500 from a newspaper ad or catalog
- Pencils or thin markers
- Chalkboard or whiteboard for teacher
- Chalk or markers for teacher

SUCCESS STEP: Have the student read a price in dollars (with no cents) that you have written (e.g., $5.00) on a chalkboard or whiteboard. Make sure the price is one that the student can read easily.

PROCEDURE:
1. Show your student(s) the Reading Prices Worksheet with dollars and cents prices. If possible, use items and prices for something that you know the student enjoys. Read one of the prices and write it on the worksheet. Then both you and your student should read that price out loud.

HUNDREDS	TENS	ONES	DECIMAL POINT WALL	CENTS	
	read	together	•	read together	
4	2	3	•	5	5
Four Hundred	Twenty	Three	AND	Fifty	Five

2. Review the principles of correctly reading the sample price: saying the dollar amounts, saying *and* for the decimal point, and reading the two places of the cents correctly.
3. On a separate piece of paper, have each student write the prices in words on the Reading Prices Worksheet. If the student is unable to write the words for the numbers, he may read them out loud to an adult.
4. Using their worksheets, have all the students read the prices out loud.
5. Regularly have the students read prices from the newspaper or store windows to maintain that skill.

Kyle was writing down numbers as his sister called them out. She read out, "267." He wrote down 20067. He was writing down exactly what he heard her say, but he ended up writing twenty thousand sixty-seven That is one reason that in Teaching Math, Book 1, we practiced with Fat Mats (place value tables) that are already labeled with the names of the units. Kyle's teacher had him do a couple of Fat Mats to review. In addition, his older sister told him, "You don't write the hundred. It is just the name of the place, silly." And that took care of the problem.

CHECK, CHECK GAME

OBJECTIVE: The student will be able to accurately read dollar and cents prices.

MATERIALS:
- Die
- Blank checks (6 for each) from Appendix B
- Pencils

PROCEDURE:
1. Each student in turn throws the die. After the throw, the player writes that number on one of the underlined blank spaces on the first check on the game sheet.
2. The players take turns throwing the die 4 times each. Each time the student writes the number on one of the underlined blank spaces on the check. After the fourth throw, each player should have a check with dollars and cents.
3. Each student reads his check amount out loud. The person who has the largest amount on his check is the winner for round 1. If the person does not read the price correctly, his check doesn't count.
4. The players continue playing for 5 rounds. The person who wins the most rounds wins the Check, Check game.
5. For more advanced players, you could make 5 blank spaces and roll the die 5 times.

The Calculator and Money

A problem with using a conventional calculator to compute money sums is that the calculator will drop the zero in the units place. (For example, if the answer comes out as $5.20, the calculator will show 5.2.) This makes it difficult for many students to understand what the answer is in dollars and cents. A calculator called Money Calc from PCI has a setting called "money" that will always show two places. (This calculator also has a setting for figuring tax and tips.) This calculator can be a good option for students in elementary school, but it is brightly colored so is not as suitable for secondary students. A better choice for older students is the Texas Instruments model TI-15, which can be set to show two digits to the right of the decimal point as needed for money.

The activity below involves completing a money worksheet where some of the answers will show only one digit to the right of the decimal point if solved using a conventional calculator. If the students have difficulty in expressing these answers as dollars and cents, I would suggest that you buy one of the above calculators and use it until the student is able to understand the place value of money as shown by the common calculator.

CALCULATING MONEY

OBJECTIVE: The student will be able to add and subtract dollar and cents amounts on a calculator and express cents correctly even when shown as one digit.

MATERIALS:
- Calculator Money worksheet
- Paper and pencils
- Slate or whiteboard
- Calculators for each student

SUCCESS STEP: Ask one student to write $1.25 + $1.25 on the slate. Help, if necessary.

PROCEDURE:
1. Show the problem of $1.25 + $1.25 = $2.50 on the board. Do the problem on the calculator. On the regular calculator the answer should say, $2.5. Ask them how much this is. If they say "two dollars and five cents" ask them how two dollars and five cents SHOULD be written—hopefully, they'll say "2.05." Then you can point out that $2.5 CAN'T be the same as $2.05 because the 5 is too close to the decimal. If they say $2.5 is $25 dollars, tell them to look again—why is that decimal point in there if it's $25? Tell them that in money you always have to have two places to the right of the decimal point.
2. Go through the first three problems with the students.
3. Have them do the rest of the problems independently.
4. Check each problem to see if the students have done the problems correctly.
5. Explain each problem, if necessary.
6. Correct the worksheets and decide whether to continue with the regular calculators or to use calculators that can be set to two decimal places or the Money Calc.

Next-Highest-Dollar Strategy

Although we advocate working with currency before we work with coins, there is no escaping that the prices in advertisements and stores promotions use both dollars and cents. So, although survival math does not require students to be able to count out exact change for purchases, they do need to be able to understand what cents are and that cents appearing after the decimal point increase the price. An important strategy for working with money is the skill of rounding off the prices to the next highest dollar and then giving the cashier more than the stated price. This is such a useful skill that it was introduced in *Teaching Math, Book 1* as a money survival skill. *Book 2* will offer more practice so the student can use the strategy almost automatically.

TALKING THROUGH THE NEXT-HIGHEST-DOLLAR STRATEGY

OBJECTIVE: The student will be able to give the cashier or teacher one more dollar than the stated price of several items.

MATERIALS:
- Slate or whiteboard and writing materials
- Next-Highest-Dollar worksheet from Appendix B
- Clothing, music, video, or other ads from the newspaper (not food ads)
- Realistic play currency

SUCCESS STEP: Have the student point to an advertisement in the newspaper or catalog of something he would like to get for himself. See if he can read the price correctly. Praise his selection.

PROCEDURE:

1. Point to the ad of an interesting item (with both dollars and cents) and ask the students how many dollars (say you have no coins) it would take to buy it. You will probably be able to tell from the students' answers if they already understand and use the next-highest-dollar strategy.
2. Explain that if you don't have coins, then you have to pay more than the price, and the cashier will give you cash back. Have the students look at the number directly to the left of the decimal. Then have them count one number higher to find the next highest dollar.
3. Go through the ads you have brought that have dollars and cents listed and have the students tell you how many dollars they would give the cashier. Don't use ads that advertise 2 for 1 sales or quantities on sale. Just show the students the prices that have a picture and a price with dollars and cents.
4. Have the students do the exercise on Next-Highest Dollar from Appendix B. Fill in the 3 blanks with some items you think the student would actually like to buy.
5. You can make this activity more hands on by having the students count out play money to "pay" for the item they want.

SHOPPING FOR ONE ITEM WITH THE NEXT-HIGHEST DOLLAR STRATEGY

The most important way to teach the next-highest-dollar strategy is to have your student save for some item that he really wants and when he has enough money, have him go to the store and buy it. If he has already indicated that he wants an item such as a CD, that is an ideal time to teach him about saving for it and then buying it himself. If he has not expressed an interest in buying anything, ask him about something that he would like and perhaps give him some suggestions. Saving for two weeks would be a long enough period of time for this lesson. Parents can offer to pay half of the cost if it would take their child a long time to save for it.

Have the student write down (or have someone write for him) the item he wants to buy. Then have him look in ads or catalogs to find a price that he may have to pay. If possible, have him bring the ad when he goes shopping. If there is sales tax on that item, tell the student he will have to have one or two more dollars to "pay the governor." In the next chapter, students will be taught how to figure an approximate tax.

Note: Explain that sometimes when you use the next-highest-dollar strategy, you have to give the clerk more than just one dollar more. For example, an item might cost $6.44, but the student has only a ten-dollar bill. The next highest dollar would be $7.00 However, the student would have to give the clerk $10 and the clerk would give him change.

Parents will find many occasions to exercise their child's skill of rounding up to the next dollar. Teachers will probably have to arrange for a community trip or use the school store, if the school has one. It is important to let the student take the lead and only help if asked—or if a *major* error is going to occur. If you are taking several students, it would probably be wise to choose a large store such as Wal-Mart that has many items for sale so the students can all shop in one store. The pictures above and on page 198 show a small group that a parent volunteer and I took to the store to buy just one item. (One of the students had enough money for three items, which he proceeded to buy.)

SHOPPING FOR MULTIPLE ITEMS WITH THE NEXT-HIGHEST-DOLLAR STRATEGY

Using the above procedures, have the student go shopping again, but this time buy two or three items. The student then needs to use a calculator and add up the cost of all the items and figure the next-highest dollar on the total amount. At first, it might be easier to make sure that each student has several one-dollar bills so he can buy with the next highest dollar. However, most students can quickly figure out the closest amount they can make with their existing bills, even if that means overpaying by $5 or more.

Watch how each student handles the entire shopping situation. If possible, go to the store when there are few other shoppers so the students can take time with their purchases. Help the students handle putting money back in their wallets after paying the cashier.

For parents, it would be valuable to repeat this experience as much as possible so the student can do it easily. Consider going to different stores when the student masters the process at one store.

EARN AND PAY GAME
(STARTING WITH LARGE DENOMINATION CURRENCY)

The game is played just as before. See Appendix B for game. This time, however, give the students only ten- and twenty-dollar bills to start out with. Therefore, they must pay with the next-highest-dollar strategy or lose a turn. In this case, if the student has only twenty-dollar bills, he may have to go higher than the next highest dollar. For example, he may have to buy something that is $17, but he has no ones or fives. He will have to give the banker the $20 and expect more change. By the time students have reached this level, they are usually ready to do the subtraction (by calculator) to figure out how much change they should get back.

Of course, if the amount exactly matches the currency the student has, he does not have to go one dollar further.

Season was accurate in figuring out her money when she wanted to buy some hair barrettes. She handed the five dollars to the cashier, but when the cashier handed back her change, she dropped everything on the floor. When her mother asked what happened, she said, "I didn't think about getting the change back. I closed my purse. The next person in line was getting close. I just grabbed the money and my purse and hurried past the cashier. When she handed me my package, I just dropped everything."

Season had not been prepared for the actual task of buying items. She forgot about getting change back, and she didn't take time to put her money in the wallet in her purse. Having someone waiting made her feel that she had to hurry. She really had too many things in her hands, which resulted in her dropping them. Her mother rehearsed the exchange with the cashier at home several times. She also taught Season to look for a part of the counter or a nearby counter where she could stand and put away her money carefully.

Games to Buy or Try

- **Moneywise Kids** (TaliCor). This is a fairly simple game with dice and play money in which players add or subtract small amounts to "earn" $100 or spend $100.

- **Make Your Own Opoly** (available on Amazon.com). You can make your own game similar to Monopoly, but you choose what is on the board and what the prices are.

- **Money Dominoes** (Learning Resources). Oversized cardboard dominoes show pictures of currency and/or coins on one end of the domino and written amounts on the other. The point is to match the pictures with the correct amount.

Coins

Your student(s) should have learned the names of the coins and their values when working on *Teaching Math, Book 1*. The following activities offer a brief review. However, remember that understanding the value of coins is more difficult than understanding the value of currency. First, their relative sizes don't match their relative worth. For instance, the nickel is heavier and bigger than the dime, although the dime is worth twice as much. Second, the appearance of coins changes frequently. Newer nickels have an updated picture of Jefferson on the front and a variety of scenes on the back. Quarters now have one consistent picture on the heads side, but a variety of different pictures representing the various states on the other side. Soon there will be 50 different pictures on the tails side of the quarter. And what is often frustrating, there are so many combinations of those coins that can equal one dollar.

After the student has reviewed the names and values of the commonly used coins, a section on skip counting coins will be included as an optional activity.

IDENTIFYING AND VALUING COINS

OBJECTIVE: The student will be able to name and tell the value of nickels, dimes, and quarters.

MATERIALS:
- One (real) quarter, nickel, and dime for each student in a plastic bag
- 20 squares of aluminum foil about twice the size of a quarter for each student
- Index cards with real coins taped on them and their value (3-4 each of nickels, dimes, and quarters) for the teacher to test the students at the end of the lesson)

SUCCESS STEP: Have the student identify a quarter out of his supply of coins. Praise his correct identification.

PROCEDURE:
1. See if the students can identify the real coins. If they need more instruction, use the Money Chapter in *Teaching Math, Book 1*. You may want to review the coin mnemonics discussed in Book 1:

> *Penny, penny*
> *Easily spent*
> *Copper brown*
> *And worth one cent.*
>
> *Nickel, nickel,*
> *Thick and fat.*
> *You're worth five cents,*
> *I know that.*
>
> *Dime, dime*
> *Little and thin.*

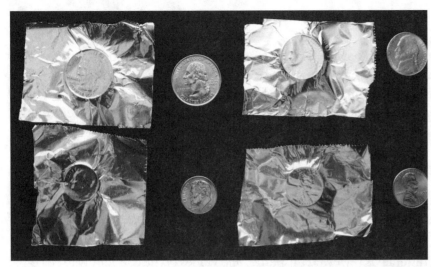

I remember
You're worth ten.

Quarter, Quarter
Big and bold.
You're worth 25
I am told.

2. Have the students make coin rubbings of each of the coins by placing the foil squares over the coins and rubbing their fingers on the foil. Have them make rubbings of both sides of the coin.

3. Give students the three index cards that have a coin taped on them and the appropriate value.

4. Mix up the foil squares representing the coins in front of each student. Set up the index cards in a line. Have the student put the foil squares under the appropriate index card.

5. To sort the coin rubbings, the students will need to pay attention to the details of the coins and get a tactile impression of them. Actually, it is more difficult to recognize the coin rubbings than the actual coins because you don't have the different weights to help make the distinction.

6. Later, turn over the index cards used for matching above, and mark the value of the coin on each. See if the students can match their coin rubbings with the appropriate value without having the coin as a model.

7. Have each student sort real change using the index cards with the number value only.

Using Coins in Vending Machines

The major use for coins in our society is for vending machines. Vending machines are available in many settings and are often used by all students. Of course, students should ideally know whether they have the correct change for the item in the machine. However, that is not always necessary for everyday use. Most machines give change if you put in too much money. In addition, many machines show a running tally of how much money you have deposited. Therefore, a variation on the next-highest-dollar strategy can be used. Just put in more coins than the price indicated and get the item and change back.

A couple of years ago, my son explained to me how he got items from the vending machine. Mostly, he used quarters. He knows that 4 quarters make a dollar, so any item priced under a dollar can be bought with 3 or 4 quarters. If it costs more than a dollar, he puts another quarter in. "I usually get money back," he said. I asked him what he does when the vending machine says exact change only. "I just put the money in and push the button each time until the snack comes out." I told him he could be losing some money that way. "I know—but it is not much money—only coins, and I

usually really want the drink or bag of chips." I wasn't too happy with that strategy at first, but after thinking about it, I realized that it was quite adaptive and worked for him. I asked him *now* if he still does vending machines that way. "Oh, Mom," he said, "I know what the coins mean and I can do it in my head." We had never worked on that skill, so he had been able to progress to the more conventional use of coins in the vending machines without direct instruction.

In order to use coins in vending machines the conventional way, students have to know how to count change, which is an optional skill that is taught at the end of this chapter. Give the student a lot of practice using the vending machines and make your own judgment about whether he can learn the standard way to count change for the vending machine. It is usually easy to keep the student motivated to learn vending purchasing skills because he gets a built-in reward at the end.

Frequently, in special education classrooms, the student is given a pre-selected amount of money for one item from the vending machine. He is then taught how to operate the machine. The real skill, however, is knowing how to figure out whether you have enough money and then operating the machine. Students should have many occasions to use this skill—at home, at school, and in the community. It is almost as useful as knowing how to order at McDonald's!

While Emma was waiting for her sister to finish up ballet class, she kept nagging her mother for money to use in the vending machine. Finally, her mother gave her three quarters. Emma looked at the coins and said, "OK, seventy-five cents" and went off to visit the vending machine. About five minutes later, she came back looking very disappointed. "I can't buy anything!" she complained. "Nothing costs seventy-five cents." Her mother went with her to check out the vending machine situation. Sure enough, nothing in the machine cost exactly seventy-five cents, but there were items inside that cost $0.55, $0.60, and $0.70. Although Emma knew how to count coins, she did not know that she could buy anything from a vending machine that cost less than the amount of money that she had—that she could get change back.

Counting Coin Change (Optional)

Counting coins as change is a complex task, and some authorities feel that it may not be a good use of instructional time for students with disabilities (Browder & Wilson, 2001; Browder & Snell, 1993). If you think about it, very few individuals without disabilities stop to recount their change before putting it away after a purchase. And, although it is possible that an unscrupulous cashier might try to shortchange a

student with disabilities, the maximum amount anyone could be shortchanged with coins is $0.99. (Hopefully, students who are taught money skills with this book will realize if they do not receive enough *currency* in change and speak up.)

Students who are in inclusive settings in the lower grades may need or want to work on coin counting skills at the same time their classmates do. Coin counting is not usually taught in the general education classroom beyond about the second or third grade, however. If learning to count coins is determined to be an important goal for an older student, that goal may have to be worked on at home or in a special education classroom. Regardless of the student's age, you may want to try a method that has been successful for some hands-on learners. It involves counting by fives as was done with currency. This is an optional skill that you may try with the student but not teach to the point of frustration.

More than one method has been proposed to skip count coins. Touch Math (Innovative Learning Concepts) uses a skip-counting scheme to count coins. Students are taught to visualize dots on the coins and count by fives. Touch Money (PCI) also teaches students to skip count fives while picturing 1 to 5 touch points on a coin. Another method developed by Margaret Lowe and Anthony Cuvo (1976) teaches students to skip count by fives with the aid of finger cues. The method used here is probably closest to the Touch Money technique. If a student can already skip count ten-dollar bills competently counting by tens, he may not need to use the method discussed below. Instead, help him understand that he can count dimes the same way he counts ten-dollar bills.

COUNTING NICKELS, DIMES, AND QUARTERS

OBJECTIVE: The student will be able to count nickels, dimes, and quarters using the touch five method.

MATERIALS:
- Large pictures of a nickel, dime, and quarter from Appendix B
- Small coin pictures with dots from Appendix B
- Real coins for each student
- Red fine line markers for each student
- Nickels Worksheet, Dimes Worksheet, and Dimes and Nickels Worksheet from Appendix B for each student.

SUCCESS STEP: Have the student skip count by 5s to 25. If he is not able to skip count fives accurately to 25, you model counting and have the student repeat. (If the student is not able to count by fives at all, go back to the section earlier in this book on counting by 5s and review them before continuing.)

PROCEDURE:

1. Have the students practice skip counting by 5s to 100. You can help them by chanting or singing the fives. See *Teaching Math, Book 1* if you need to reteach skip counting. You may want to use the small book, *Reese's Pieces, Count by Fives* (Pallotta, 2000). If the students have been successful with counting currency by fives, they will probably be ready to use the touch coin method.
2. Introduce the large pictures of a nickel, dime, and quarter from Appendix B.
3. Using the pictures of the coins, mark the dots (indicated by a circle) with a red marker, as shown below. Touching the red dots, skip count by fives as a model for the students.

4. Have the students put red dots on the small coin pictures from Appendix B.
5. Have the students count (aloud) the touch points (red dots) on the pictures.
Nickel—1 dot = 5; dime—2 dots = 5,10; quarter—5 dots = 5,10, 15, 20, 25.
6. Using the red marker, have the students put dots on the real coins to match the pictures.
7. Then have the students skip count the nickels (5, 10, 15, etc.) in the Nickels Worksheet from Appendix B.
8. When the students are able to skip count the nickels, introduce counting the dimes by saying 5, 10 on each dot in the Dimes Worksheet from Appendix B.
9. When the students are secure with dimes and nickels, have them do the Dimes and Nickels Worksheet from Appendix B.
10. Give the students plenty of practice counting real nickels and dimes (or realistic play coins).

An Optional Method of Teaching Skip Counting

Counting by fives is such an important skill that you may want to use another method of teaching skip counting from 5 to 100, if the student is not getting it. Flap charts can allow a student to quiz himself on a variety of important facts and concepts. To make one, fold a piece of cardstock or construction paper in half. Cut the top half from the bottom of the paper to the fold to form 5 flaps that can be opened to see the back paper. You may want to staple each flap at the top. (You can also use the top half of a manila folder and just cut strips in the top half of the folder, leaving the back of the folder intact.) On the back paper, write the 5s: 5, 10, 15, 20, 25.

The student studies by looking at the number and saying it, putting the flap down so the number doesn't show, and saying the number. The student then practices doing more than one number until he can say all the numbers without assistance. The teacher then makes a second flap chart with the numbers 30, 35, 40, 45, 50, etc. This is somewhat similar to using flashcards, but the student can self-correct and go at his own pace. See page 176 for an illustration of a flap chart used to teach units of measurement.

COUNTING NICKELS, DIMES, AND QUARTERS

OBJECTIVE: The student will be able to count nickels, dimes, and quarters using the touch five method.

MATERIALS:
- Large pictures of a nickel, dime, and quarter from Appendix B
- Real coins for each student
- Red fine line markers for each student
- Mixed Coin Worksheet from Appendix B

SUCCESS STEP: Have the student skip count 3 dimes by fives. Praise his success.

PROCEDURE:
1. Then have the students count the quarters in the picture of quarters, counting the five dots —5, 10, 15, 20, and 25.
2. Have the students do the Mixed Coin Worksheet (nickels, dimes, and quarter) from Appendix B.
3. Have the students count real coins to each other in pairs. Have each student skip count a variety of change to the teacher.
4. You may want to teach half-dollars using the same method as above; however, it is unwieldy to count by fives ten times to get to 50. Since half-dollars are used very little, you may just teach the student the value for the half-dollar and that two of them make up one dollar.
5. Since quarters are used so frequently, especially in vending, you may want to teach the students to count quarters to a dollar by using a chant:

> *25, 50, 75, a dollar.*
> *We know our quarters*
> *So let's give a holler.*

6. Another mnemonic that might be useful when counting mixed coins involving quarters is:

> *Take one quarter, add one dime,*
> *It's 35, every time.*

An Alternative to the Touch Five Method

Some students are more able to skip count coins by fives if they do not touch dots on the coins, but point with one finger toward the nickel for 5, two fingers for a dime, and all 5 fingers for a quarter (Lowe and Cuvo, 1976). Have the student try the finger cueing method if he has difficulty keeping track of the number of dots that should be counted on each coin.

COIN BINGO

OBJECTIVE: The student will practice adding two coins of different denominations (later, three coins).

MATERIALS:
- A brown paper lunch bag or other opaque bag
- 10-15 each nickels, dimes, and quarters (real coins, if possible)
- A bingo board for each player, from Appendix B—or make your own by drawing a grid with 9 or 16 squares on a sheet of paper, cardstock, or cardboard. In each square, write (in random order) one of the amounts that can be made by adding two coins together: $0.10, $0.15, $0.20, $0.30, $0.35, $0.50
- Pennies, magnetic bingo chips, or other tokens—10 or 12 per player

PROCEDURE:
1. Mix up all the nickels, dimes, and quarters in the bag.
2. One player reaches into a bag and takes out 2 coins. He says the total, and he and the other players use a token to mark that amount on their boards, if they have it.
3. The next player reaches into the bag and draws two coins and announces the total. Again, all players cover that space on their boards. (In this case, it's OK if players kind of "cheat" by feeling around in the bag for coin combinations that they need on their board. It's good practice!)
4. Players continue passing the bag, drawing two coins, and announcing the total. The first player to get three tokens in a line, vertically, horizontally, or diagonally yells out "Bingo" and is the winner.
5. For more advanced players, use the bingo boards showing amounts you can make by adding THREE coins (nickels, dimes, and quarters): $0.15, $0.20, $0.25, $0.30, $0.35, $0.40, $0.45, $0.55, $0.60, $0.75, and play as above.

ADDING PENNIES

OBJECTIVE: The student will be able to count coins, including pennies.

MATERIALS:
- 10 pennies for each student
- Mixed Coin Worksheet (Including Pennies) from Appendix B

SUCCESS STEP: Ask the student to give you three cents from the pennies on the table. Praise his successful response. If he is incorrect, remind him that cents and pennies mean the same thing and repeat the request.

PROCEDURE:
1. Show a penny to the students. Tell them that when you have finished counting the silver coins, you add the pennies by making the answer 1 more for each penny. Practice adding 1 or 2 pennies to the problems you did with the silver coins.

2. Practice counting a variety of real coins, including pennies. Make sure you give the students practice counting up coins that total $1.00, so you can make sure they know that after you get to $0.99, if you add on one more cent, you say "one dollar" rather than "one hundred cents."

3. Have the students do the Mixed Coin Worksheet (Including Pennies).

If you have a student who cannot get the hang of counting coins, remember: Being able to count coin change is not really a survival skill. Most prices are more than $1.00, and the student who can use the next-highest-dollar strategy can manage quite well without knowing exactly how much change he has in coins.

GENERALIZATION ACTIVITIES

Ann's family had a habit of dropping change into a large jar. Usually they waited until the large jar was full, then took it to a coin counting machine at

the bank. That machine gave back currency and a few coins for the total amount of money. To give their daughter experience counting change, they used a smaller jar and emptied it every week. Ann sorted the coins and used skip-counting coins to figure out the total each week. Then she took it to the coin counting machine to see if she was right.

1. One father made a habit of emptying the change out of his pockets each day. His son really enjoyed counting his father's change. Because the father usually had a lot of change, they worked out a method of putting the coins in piles that added up to one dollar, so the son could skip count the piles as if adding $1.00 bills.

2. Encourage your child to save his change in a bank or jar. The banks that keep a running total of the amount you put in can be educational if you encourage your child to observe how much the total increases for each coin added. You can also use an old fashioned bank and periodically count up the total. With either type of bank, you can then take the coins to a coin counting machine or wrap them in coin rolls to take to the bank.

3. On road trips where you pass through tollbooths, ask your child to give you the appropriate amount of change as you approach the booth.

4. Before you ride the bus or subway, have your child count out the correct change for his fare. If he is older and often rides on public transportation, have him learn all the different coin combinations to equal his fare.

5. Ask your child to read prices when he asks for something in the store. "How much are those cookies?"

6. On another occasion when your child asks for something at the store, tell him you will buy it if he can find a package for less than $2.00 or another price you choose.

7. At school, collect coins to donate to a worthy cause or to buy supplies for a party. Have the students total them up before taking to a bank or coin counting machine.

CHAPTER 14

Shopping

Questions to be answered:

Can the student:

1. Find the cost of one item from pricing for multiples.

2. Find the cost of items priced per pound.

3. Add up 3 or 4 items correctly when the prices have both dollars and cents (can use calculator).

4. Find the unit price on the label of appropriate items and compare two like items to find the best buy.

5. Figure state tax and other charges.

One of the major uses of math in real life is for shopping. We not only use our computational and money skills to understand financial issues such as how much things cost and whether we can afford them, but also to resolve psychological issues such as whether we should or should not buy something, given our wants versus our needs.

This chapter will focus first on shopping for food because that is essential for most people. The way food is advertised and priced may make food shopping more difficult than other types of shopping. For example, food prices are often advertised as several items for a certain amount, such as 2 for $1.00. How does the student know how much one item will cost? And meat and produce are often priced per pound. How does the student know how much the meat for a meal will cost? It can also be difficult for students (and adults) to comparison shop because of the different ways items are priced. For instance, which is cheaper? 20 ounces of pasta for $1.59 or 1 pound of pasta for $1.00? Students need to learn to look at labels (or at unit pricing signs on the shelves) to learn about unit pricing.

Figuring sales tax, which is important when shopping for most items, will also be discussed in this chapter.

Grocery Shopping

Students usually need lots of practice adding several prices up to get the total spent. In addition to the worksheets that are provided here, most parents and teachers could use one of the workbooks that are sold in bookstores, grocery stores, discount stores, variety stores, etc. if they find the student needs extra practice. Parents can also get their child to practice this skill in real life, such as by having her add up the costs of one category of food products (dairy, vegetables) when they are doing the regular shopping.

PRACTICE IN ADDING DOLLARS AND CENTS

OBJECTIVE: The student will be able to add up the prices of 3 or 4 items correctly when the prices have both dollars and cents.

MATERIALS:
- Totaling Your Shopping List, Appendix B
- Pencils
- Calculators

SUCCESS STEP: Ask the student to read some of the prices of the items on the Shopping List worksheet. Praise correct answers.

PROCEDURE:
1. Have the students point to the first, second, and third item along with their prices on the worksheet.
2. Have them carefully enter the numbers into the calculator, emphasizing the need to put in the decimal point accurately.
3. Check the answers.
4. Repeat above with second problem and check.
5. If the students demonstrate understanding, assign them one or two problems to do independently.

6. If necessary, reinforce that when the calculator rounds off a dollar amount to just a whole number, like 1 or 5, that means $1.00 and $5.00, and when the calculator changes a money amount so there is only one number after the decimal, they need to mentally put a zero at the end (e.g., $2.5 equals $2.50). See previous chapter on money for more detail.

FIGURING THE COST OF ONE ITEM WITH PRICING FOR MULTIPLE ITEMS

One of the immediate frustrations of shopping for food is the way products are advertised and priced. Many ads list the prices as 2 or 3 for a certain price. We can teach the students how to divide to get the price for one item—but we don't always know if the store will let them buy one item at the price they have calculated. The store may be having a sale to push quantities of that item and will charge a higher price if the customer wants only one item. If that is the case, the ad will usually give the unit price and the price for when you buy multiples such as $.40 each or 3 for $1.00. Students must be taught to carefully read the ad or price—including the small print. When I took the group of young adults I was teaching to do food shopping, I found that most of the prices for one item were pretty close to the amount we had calculated by division.

OBJECTIVE: The student will be able to figure out the price of one item when the price is given for two or more items (assuming the store has marked the item proportionally).

MATERIALS:
- Food ads from newspaper or flyers
- Two food items that are shown in the ads and are priced as 2 for $2.00. For example, two cans of peaches for $2.00, two small boxes of strawberries for $2.00, or two pens for $2.00. (If you can't find a food ad that is appropriate, make up your own food flyer advertising the two food items that you actually have for $2.00.)
- Two single dollar bills (real or fake)
- Calculators
- Pencil and paper
- Shopping for One Worksheet (Appendix B)

SUCCESS STEP: Have the student identify the pictures of the foods and other items given on the ad. Then have her read the prices from the ad. Praise success or correct and have student repeat the correct labels.

PROCEDURE:
1. Point to one of the items that is advertised at 2 for $2.00 (or 3 for $3.00, etc.). Ask the students what the price would be if you only wanted to buy one item.
2. Show the students the portion of the ad that advertises the two items for $2.00. Hold up the two single dollar bills and ask the students if you have enough money to buy the two food items (yes).

3. Then ask the students what to do if you only want one of the items. They might be able to tell you that one of the items will cost $1.00. If not, you can help them see that one item must be $1.00 by laying one bill on each item and saying "One dollar, two dollars."

4. Check the answer by calculator. Say, "The ad says 2 for $2.00. We want to share the money evenly for each identical item, so we need to divide. First put the price in the calculator. Then divide by the number of items (2) to get the price of one item."

5. Have each student do the problem on her calculator. Show them that the answer is the same as when you dealt out the real dollars, one per item.

Note: If the student has a difficult time with the calculator omitting the zeros from the right of the decimal point (2.00 ÷ 2 = 1), you may use a Money Calc (PCI), which has a money setting, or a calculator such as the Texas Instrument TI-15, where you can set the answer to show 2 places to the right of the decimal point. You may also teach the student that there must always be two places (dimes and cents) to the right of the decimal point when you are working with money. "The calculator isn't smart enough to remember that, so you have to check and add a zero or two if necessary."

6. Explain to the students that *per* means that the total number or price is going to be divided by the number of items. They can use that algorithm (way to solve the problem) when the problem says *per* or *for*.

7. Have the students work the first two items on the Shopping for One worksheet from Appendix B. Correct those two problems together with the students. If they seem to understand the process, let them do the rest of the worksheet independently. If you see that a student seems to be just guessing, quickly give her individual help so she does not practice the skill incorrectly.

Jake saw a sale sign that said, "Subs – 2 for $5." He divided 2 into $5 and got the price of one submarine sandwich as $2.50. When he ordered the sub, the cashier asked him for $3.00. He protested that the price for one sub was $2.50. The cashier said, "The price is only good if you buy 2 submarine sandwiches. The price for one sandwich is $3.00." In that area of the country, when buying nonfood items or food in a restaurant, the sale price only applies if you buy the number specified on the sign or in the ad.

You will have to determine what is customary in your area. Often the ad or sign will give the one-item price in smaller print.

GENERALIZATION ACTIVITIES

The best way to generalize this skill is to have the student go to the store and figure out the price for one item when the price is given for two or more. Parents can usually just take the student to the grocery store and have her use her calculator to figure all the unit prices for items that the family actually buys.

At school, students can figure the price for one item at a school store, a pretend store in the classroom, or at a bake sale. Teachers could also tell the class the total price for their whole group to do something (like take a field trip) and then have them divide the total amount to see what each individual student will have to pay. Teachers should alert the parents to how they can help their student learn about shopping for groceries because schools just cannot provide enough real experiences to make the teaching permanent.

Parents, when grocery shopping with your child, here are some special situations to point out as she seems ready to understand these concepts:

- If something is advertised as being "2 for the price of 1" that is not the same as 2 for $1.00. It means you can get 2 items for the same price you would ordinarily pay for 1 item.
- Just because something is advertised as 2, 3, 4, etc. for a given price doesn't mean you cannot buy *more* than that number of items. For example, if lemons are 2 for $1.00 and you want to buy 3, you still begin by finding the price of one by dividing $1.00 by 2. Once you have the price of 1 item, you multiply that price by the number of items you are buying.

PRICING PER POUND

Moira wanted to buy one banana for each person in her six-member family. When she went in the grocery store, she started to cry. "I don't know if I have enough money. The sign says $ .89 per lb." Her mother had to help her weigh the 6 bananas and then multiply the weight by $.89. Moira got the idea, but her mother had to figure out what 3 lb. 4 oz. would be in pounds (3¼).

Until recently, most grocery stores in the U.S. had scales for customers to use to weigh their fruits and vegetables so they could roughly figure out the cost of their produce. Now some stores do not provide a scale for customers because they believe that not as many customers want to use one. Some store personnel also feel that the scale used for pricing is so much more accurate than the customer scale that the customers would be confused. Some stores now sell all produce in pre-weighed packages which simplifies knowing how much you will pay for a given number of fruits.

Another problem related to shopping for fruits and vegetables is that customers often do not know how much produce is needed for the number of people to be fed. I usually estimate that two big fruits such as apples equal one pound, and four little fruits such as plums weigh a pound. However, the size of fruits varies and changes the estimates. If the store has a customer scale, the students can actually weigh some of the fruits to see what one fruit weighs. Students can be taught to compare the prices per pound to decide what is the best buy. (However, I have found that their choices are often made more on their likes and their dislikes than on the best buy financially.) Vegetables are also often priced per pound, but they don't seem to motivate the students as much as fruit.

Meat is usually the most expensive portion of the food budget. Here again we can compare price per pound, but there are other factors, such as whether the bone is left in or how much fat there is, that affect how much meat can actually be used. Students can best learn about these factors by looking at the meat in the store and then again after it is cooked. Parents really have an advantage over teachers in helping their children learn these shopping skills since they have many more opportunities to teach about them in the grocery store.

Four researchers recently determined that the most effective way for students to learn purchasing skills is through a combination of simulated and real-life instruction (Y.P. Xin et al., 2005). Unless someone helps the student buy things over and over again in different settings and situations in the community, she will not learn to use much of the money and shopping skills taught in the above procedures (simulations) in her daily life.

For this book, we will just teach how to figure out the cost of an item when it is priced per pound and leave the many details to families to teach about during real shopping trips.

CALCULATING PRICE PER POUND

OBJECTIVE: The student will be able to calculate the total cost of an item when it is priced per pound.

MATERIALS:
- A bathroom or food scale
- A bag of one kind of fruit and or vegetables weighing 2 or 3 pounds (not pre-priced)
- A package of meat or cheese that is wrapped and priced (you may need several packages)
- An ad giving prices for the fruit (price per pound)
- Calculators, paper, and pencil

SUCCESS STEP: Ask the student what the bathroom or food scale is and what it is used for. Demonstrate the scale if she is incorrect and then have her weigh something.

PROCEDURE:
1. Have the students take turns weighing the bag of fruit and meat on the scale.
2. Show the students the price tag that goes with the fruit such as $1.00 per pound.
3. You weigh the fruit and write the weight on a paper or slate that the students can see. Go to the next highest full pound for the weight. (For example: apples – 3 pounds.)
4. Ask the students how much they will have to pay for all this fruit. Guide them to say that they must multiply the price per pound times the number of pounds to get the answer.
5. Have each student do the problem on her calculator.
6. Do at least one more simple problem using more fruit or 2-3 pounds of vegetables.
7. See if you can get the students to come up with a rule such as: When the price is given per pound, multiply the number of pounds times the price per pound.
8. Show the student the wrapped piece of meat or cheese and see if she can find the price per pound and the actual weight of the item on the label. Do the problem

using the calculator (total weight times the price per pound) to see if the piece of meat (cheese) is priced correctly. You may need to show the students several pieces of meat so they can understand the labels in the store.

9. If the student is catching on well to the per pound pricing, you might try to teach her about per ounce pricing. You will have to teach her that 16 ounces is equal to one pound to make the information really useful to her.

Unit Pricing/Comparison Shopping

Luckily, we do not have to use the calculator to find the unit price for every item that is priced in multiples. Food stores usually put the unit price of the item somewhere near the item on the shelf. The label usually shows the name of the item, the total price of that package, and the unit price. If you can find the unit price on the label, you can compare the price of two items even if they are packaged in different sizes. If one carton of drinks has 16 cans for $8.00 and another carton has 6 cans for $4.00, it might be difficult to discover the best buy. However, if you see that the first carton has a unit price of $.50 per can and the other has a unit price of $.67 per can, you can tell what is the best buy. (However, it isn't a good buy if you can't use 16 cans of that beverage.)

Unit prices are also given by weight. For example, a large jug of orange juice might be $5.00, but $0.25 per ounce, whereas a smaller carton might be only $2.50, but $0.40 per ounce. The weight may be given in ounces, pounds, or grams. As long as the compared items use the same unit of weight, the student can compare the cost per unit of measure.

INTRODUCING THE CONCEPT

OBJECTIVE: The student will be able to find the unit price on the label of appropriate items and compare two like items to find the best buy.

MATERIALS:
- Copies of labels (supplied by teacher) for each student
- Pencils and calculator

SUCCESS STEP: Ask the students who is the best shopper in their home. Then ask them why that person is the best shopper. (There is no right answer to this question.) Thank them for their participation.

PROCEDURE:
1. Visit a grocery store where your child or student often shops and write down some examples of unit pricing as displayed on the shelves. Look for items the student would be interested in purchasing and choose two different items that have different unit prices. For example, write down the unit pricing for a one-pound bag of pretzels from both Brand A and Brand B and the unit pricing for 32 ounces of orange juice vs. 64 ounces of orange juice.
2. Type or write the unit prices for the pairs of items on a sheet of paper to make a worksheet.

3. Go through each unit label on the worksheet with your students. Ask the students to underline the total price and circle the unit price that is listed. Have them make sure that both products are measured in the same unit. Emphasize that the units must be the same to compare. You can't compare ounces with milliliters or pounds, only with ounces.

4. Choose two like items and ask which one is less expensive or the better buy. Help them understand that the item with the smallest unit price is the better buy.

WHICH PRICE IS BEST? GAME

The following game borrows a similar name from the CBS television show, *The Price Is Right*. Although the rules are completely different, we are relating to the enthusiasm shown on that show to intrigue the students.

OBJECTIVE: The student will be able to determine which item is the best (in price) from an array of 2 items using unit pricing and division, if needed.

MATERIALS:
- Which Price Is Best? game board from Appendix B
- Which Price is Best? worksheets (Appendix B) cut into 8 cards per worksheet.
- Answer sheet
- Game piece for each player
- Name badges for each player (optional – to make more like the TV show)

PROCEDURE:
1. Place markers for each player on the board. The adult or facilitator of the game calls the name of the first student and says enthusiastically, "Come on down." Players cannot take a card until they hear their names and "come-on-down!"
2. Players pick one card from the top of a pile of face-down cards. The student compares the unit prices by division and tells which store has the best buy. Some cards just require that she find the price on the labels and compare. If the answer requires division to find the unit price, the student must show her work on the calculator even if she can do the problem in her head. If the unit price is not required, she does not have to use the calculator.
3. If the answer is correct, the player advances one square toward the goal. If the answer is not correct, she does not move her marker.
4. The adult or an older student should consult the answer sheet before the player moves her marker.
5. A few of the cards are worth more than one square ahead. They are labeled so on the card.
6. The first one who reaches the cash register on the game board wins.

GENERALIZATION ACTIVITIES
Of course, the best way to help students generalize shopping skills is to have many varied shopping experiences. Before I went grocery shopping with my adult son and his roommates, I had them fill out a short shopping list. They wrote down each item

that was needed in one column, and I gave them an estimated cost that they wrote in a second column. The purpose of the estimated cost was to give them an idea of what that item should cost. I had found that although they had very good ideas of the costs of snacks, they had little idea of the cost of milk, bread, and other staples. (The younger students I have worked with do not even have an idea of what their snacks cost.)

After the young adults had some grocery shopping experience, I eliminated the estimated cost column. I then had them record the actual cost of the item while at the store. In a short time, they got quite proficient at recording the price on the shopping list. Before they went to the cash register, I had them add up the list of prices on their calculator to make sure they had enough money. I only had them buy 3 or 4 items at first so the total was easy to calculate. A larger copy of this grocery shopping list is in Appendix B.

Item that you want to buy	Estimated Cost	Actual Cost
Ex. 1 package of red Jello	$ .60	($ .57)
1.		
2.		
TOTAL	$_____	$_____

Here is the procedure I used when shopping with the young adult math group. I sent this sheet home with each student:

1. Please work with your parents or your supported living provider to choose three or four food items that you can purchase for your family or roommates.
2. After you choose the items, ask your parents or provider what they estimate the price for each item will be.
3. Write the estimated cost on the worksheet in the column labeled Estimated Cost.

4. Add up the estimated cost of the three or four items and see how much money you need to bring.
5. Have your parents or provider provide you with enough cash to buy the three or four items.
6. Then go to the grocery store with your calculator and worksheet. Get the three or four items at the store.
7. After you put each item in your cart, write the actual cost of each item on your sheet.
8. Before you check out, add the exact cost of the items on your calculator. Don't forget to put the decimal point in correctly.
9. Then check out and pay for the items.
10. Save your receipt to compare it with your worksheet. If your items are food, there will probably not be tax.
11. Bring your worksheet and receipt to class next week. (There will be a prize.)

Another good generalization idea is to have your student or students throw a party or bring refreshments to a party. It then makes sense to plan the menu ahead of time, buy wisely, and know how much money you have spent. If they have a limit on spending, the students will have to subtract the total spent from the balance of the limit.

When taking a group of students grocery shopping, I found that we had more items in our grocery cart than we had planned to buy. I observed that one student would just throw in items that she wanted when we were figuring out what was the best buy. Before we checked out, we had to find the proper aisle for each item and return it to the shelf. When I asked her why she had thrown other items in the cart, she said, "That's how I do it when I shop with my mom." Her mother said that she shopped from a list, but she had encouraged her daughter to put food and other items that she wanted in the shopping cart. Emily had never been asked to pay for these items.

Since we couldn't change the family's shopping behavior, we told Emily that she could not add items to the cart when she was shopping with people who were not in her family. She agreed and did quite well grocery shopping. The one time that she slipped and added several items, we had her find the correct shelves in the grocery store and return the items while the group waited for her.

Parents, if your child is in her final years of high school or getting ready to live on her own, your weekly grocery shopping should be part of her assigned jobs. Do not write up your grocery shopping list and just hand it over to her. Take one section of the list, such as dairy goods, and assign it to your child. Group the items as they are placed in the aisles of your grocery store. You should figure out the sizes you want ahead of time. You may know that you only need a small can of pineapple for that salad, but your child probably needs more explicit information than that. If you can put the number of ounces that are in that small can, that would be very helpful.

When I was working with my son on grocery shopping, I often just called cans small, regular, or large. I told my son that most regular-sized cans were like the can of beans I held up (except for soup cans, which were a little bit smaller). If I wanted the smaller size (like the pineapple can) or a larger can (like the apple juice can), I would write it on the sheet. I taught Scott about produce at a separate time because it is more complex (see section on unit pricing.) We also worked on meat separately.

One Step at a Time

Don't make the mistake that I did when I was teaching the young adults about grocery shopping. I tried to have them do the entire week's shopping list the first time they participated in grocery shopping. That was entirely too much at one time. Even though they had been grocery shopping for some time, they had just been putting the items in the cart and hoping they had enough money. Sometimes the supported living provider had to lend them some money to finish their shopping.

After the first disastrous shopping trip, I started giving them a relatively small list of items each week. However, I made sure that they had to pay attention to the sizes and prices of the items and use their calculator to figure out the total cost before they went to the cashier. We used the estimated costs to see if their choices were appropriate.

There was also preliminary planning to do. They had to plan the dinner menus first so they could get the right ingredients. Then they had to list the ingredients needed for dinners, and add the grocery items that they needed to buy every week.

When I look back at all the skills that are needed for effective grocery shopping, I can see that I expected too much too fast from these young adults.

Shopping for Nongrocery Items

Understanding about money and prices is only a part of shopping for nongrocery items in a mall or in a large department store. Students also need to know what general category the item they want to buy belongs in so they can find it. For example, is the bike tire going to be in the Sports Section of the discount store or in a Sports Store in the mall?

LET'S GO TO THE MALL GAME

Let's Go to the Mall is a game that can be used with individuals or with teams. A team could include: a younger student who is learning how to classify by looking at a picture of an object and finding the proper store to buy it in; another student who could add up the prices of the items on the shopping list; still another student who could subtract the total from the amount in the checking/debit balance.

OBJECTIVE: The student will be able to tell the general category of some common items so they can be located in the store (in the game). The student will also be able to subtract the cost of items from her total funds and keep a running balance.

MATERIALS:
- Game board for Let's Go to the Mall from Appendix B
- Shopping cards with pictures and prices of items to buy from Appendix B (Copy them onto cardstock or mount them on construction paper and cut out. You may just cut out the cards from the book if you think the game will have only light use.)
- Shopping Tally sheet for keeping track of the money from Appendix B
- Calculators and pencils
- Slips of paper that have the store name and the items that can be bought there (Let's Go to the Mall Store Key, from Appendix B)

PROCEDURE:
1. Shuffle the shopping cards and place them face down on the game board.
2. If the student has already done Chapter 15, Managing Money, and learned about debit cards, you can remind her that she will be using a debit card at the Mall. If not, you may introduce the concept of a debit card to her now. This debit card is not one that can also be used as a credit card. The student will only be able to use the amount of money that is available on the debit card.
3. Each student is going to have $200 on her debit card. That amount is written on the shopping score card on the beginning balance line.
4. The first student takes a shopping card from the pile. She looks at the picture and decides which store would have that item. She moves her game token to that store and looks at the Store Key to make sure she can buy that item there.
5. If she is correct, she keeps the shopping card and subtracts the price from her beginning balance of $200. If she goes to the wrong store, she does not subtract the price from her balance and the next person takes a turn. (If, in your judgment, it is possible to buy the item from the type of store that the student chooses, give the student credit for a correct choice, even if it is not listed on the Store Key.)
6. If a player gets the restroom or the playground card, she loses that turn.
7. Each student in turn buys items from a shopping card and uses a calculator to subtract the price from the balance in her debit card. The first game is finished when each player has had three turns. The facilitator checks the student's subtraction for correctness. The student who has the least money on her debit card (without going over the amount given) is considered the best shopper and the winner.
8. The games are intentionally short so that more players can have a chance to be the winner. You should play at least two games in one session. As long as the subtraction is done correctly, it is pure chance that determines the total amount of money a player is left with at the end of the game. (If chance is not being kind to a student, you can stack the shopping cards to her advantage when you reshuffle them. The child's feeling of success is more important than the randomness of the game.)
9. Once the students are comfortable with playing the Let's Go to the Mall game, you may want to change the criteria for winning. You tell them that the best shopper in the game is going to be the one with the most money left at the end of the game.

You may even want to wait to tell them the criteria for winning the game until the game is over.

This game can be played over and over again because the students won't have the same combination of 3 items. If they get tired of the items on the cards, you can paste catalog pictures on index cards and introduce new items to be purchased.

Sales Tax

It sometimes comes as a rude surprise even for adults when you have saved up just the right amount of money to buy an item, but have forgotten that you have to save more because of sales tax. Students need to be aware of sales tax whenever they buy something. Up to this time, we have just casually told them that they need a dollar or two more "to pay the governor." However, as one's purchases get bigger, that extra money "for the governor" really adds up and needs to be figured ahead of time.

Sales tax rates vary by state in the United States. I was giving a presentation on math to a group of parents and lamenting that I just hadn't found a way to uniformly add state sales tax onto a student's purchases. One mother said, "That's easy. Move to New Hampshire. We don't have sales tax." I have found that there are now five states that do not charge state sales tax. In the other states, rates vary from 4% to 7.25% as this book is published, and may go up in the future. Luckily, most states that have sales tax do not apply it to food.

One strategy is to mentally add on a uniform 10% for state sales tax. 10% is easy for the student to figure by just moving the decimal point over to the left one place, and at least for now, the student will have a little more money than is needed to buy the item. I have found that many young people (and older people too) are very afraid that they will not have enough money to pay for their purchases when they get up to the cashier. It is better to have a little more money than to be embarrassed by having too little.

ESTIMATING SALES TAX

OBJECTIVE: The student will be able to figure the amount of sales tax (at a uniform 10%) on her purchases and add that amount to the cost of the item.

MATERIALS:
- A few things that you think the students might like to pretend to buy such as a CD, a DVD, a small toy, some special item of clothing or a souvenir from a favorite place
- Price tags attached to each item
- Realistic currency (about $60) for each student
- More currency for the instructor to use as the bank
- Estimating Sales Tax Worksheet from Appendix B.
- Calculators
- Paper and pencil for students
- Slate or whiteboard for the teacher – chalk or marker also

SUCCESS STEP: Have the students name the desirable items you have put on the table before them and correctly say the price.

PROCEDURE:
1. Have the students point out an item that they would like to pretend to buy.
2. Tell them you need to pay extra money, called "tax," whenever you buy something that isn't food. In your state, taxes are an extra 5 cents (or whatever) for every dollar you spend. Explain that taxes are used by your state to help build schools, libraries, roads, stadiums, and other buildings and to run the government.
3. Tell them that they are going to have to figure out the sales tax and add it to the purchase price of the item.
4. Talk briefly about percent. It is a different way of writing a fraction. We are not going to teach about percent to real understanding here, but we will show them a shortcut way to figure the tax.
5. Tell them that we are going to figure an amount of tax that is larger than they will have to pay as sales tax. The shortcut method of figuring 10% is by moving the decimal point over one place to the left. Sometimes they will have to add a zero to the right so the money will read the right amount of cents (as in 10% of $23 = $2.3).
6. Illustrate on your slate or whiteboard prices of things they are familiar with and have them figure the sales tax. For example, 10% of a sweater costing $27 = $2.70 for the tax. Give them as many examples as needed to make sure of their understanding.
7. Have them do the Estimating Sales Tax worksheet from Appendix B.
8. Ask the students if the sales tax is all that they have to pay for the item (no). Remind them of the original price of the item and have them add the sales tax to that price to find the total cost of the item.
9. Using tagged items on the table in front of the students, have them figure the sales tax and add it to the original price and write it on paper.
10. Correct the total prices for this exercise when the students have finished. Then have the students choose one item they want to "buy" and pay you with their realistic currency. If they are correct, let them keep the items until the end of the math period.

Note: When I was giving a workshop in Canada, they told me that their sales tax was 14%. I didn't have a quick way to help them figure sales tax.

USING A CALCULATOR TO FIGURE SALES TAX

OBJECTIVE: If the students have calculators that have a % key, you can accurately figure the price plus the sales tax in a quick process.

MATERIALS:
- Calculator with a % key

PROCEDURE:
1. Punch in the price of the item; e.g., $25.00.
2. Punch in the operation sign for addition (you are adding the sales tax to the price.)
3. Punch in the rate of your sales tax; e.g., 7.
4. Then punch in the % sign on the calculator.
5. The amount given in the window of the calculator is the total (price plus the sales tax) that you will have to pay for that item.

Note: You can't punch in the equals sign as you usually do to get the answer. If you do (on my calculator), you will get double the right answer. You have to consider punching in the sign % is like punching in the equals sign.

If you want to teach the above calculator method, you will have to repeat the above exercise to figure the total cost of the items. It may be more difficult to explain that you are adding the cost of the sales tax because you don't really see the addition step. However, it is more accurate to use the real sales tax amount as done here.

GENERALIZATION ACTIVITIES

Of course, the best generalization is having the students go out and buy items for themselves. Start with one item. Have your student(s) plan the shopping ahead of time by looking for prices in newspaper ads, catalogs, and on the Internet. I advise having the students estimate ahead of time what the price will be, just as with food items. You may find that students living at home have very little concept of the prices of even common items.

I have students write their list on a small form (see below). I tell them that if you buy more than one item, the store usually figures the tax on the whole amount of what you buy, not on each item. Therefore, you could add up the total cost (real) of what you buy and only have to figure the sales tax one time

Item	Estimated Cost	Real Cost
Example: See-through backpack	$16.00	$14.50
1.		
2.		
3.		
Tax (____%)		
Total		

Some ideas for students who are not yet independently shopping:

1. When grocery shopping with your child and you are choosing between several items, ask her to find the cereal, cookies, juice, shampoo, laundry detergent, etc. with the lowest unit price—i.e., the "cheapest one." (If there are too many to choose from, tell her to choose among Brand A, B, and C.)

2. When ordering at a restaurant where there are "combo meals," figure out with your child whether these are good deals. Do you save any money if you buy the cheeseburger, large fries, and milkshake as a combo? Or is it cheaper if you buy a cheeseburger, small fries, and Coke separately? (If your child is too hungry to wait to order while you both figure this out, go ahead and order. Then, when you're done eating, figure it out so you will know the next time you go!)

3. When you see items that are advertised as being "buy one, get one free" or "buy one, get one for $1," take the time to explain how you figure out the price for one item. How do you decide if the offer is a good deal?

CHAPTER 15

Managing Money

Questions to be answered:

Can the student:

1. Record small amounts of money spent.

2. Keep track of the balance when recording spending.

3. Plan small expenses weekly.

4. Budget earnings and expenses weekly.

5. Open a bank account.

6. Fill out a check accurately and describe the purpose of a check.

7. Use a debit (ATM) card.

8. Understand how credit works.

9. Simulate the use of credit.

Managing money can be a problem even for those of us who do not have a disability (at least that we know of). Keeping track of where our money goes, planning our spending, budgeting, using credit, and banking are survival skills for us—and for students who are concrete, hands-on learners.

When I first started to help young people with handling money, I tried to teach them everything all at once. It didn't work! I also found out that managing money has as much to do with attitude and experience as it does with having the right financial skills. I needed to back up and give the students some positive experiences just keeping track of their spending and then talk about managing their money. This is the approach that I take in this chapter.

Recording Spending

It is best to focus on teaching students to keep track of discretionary spending at first. The student should have some money to spend on nonessential items. Ideally, parents can give the student an allowance and allow him to spend it on nonessential items. In school, the teacher may want to set up a play money situation where the students earn money for work or good behavior and have a chance to buy small items for themselves, or the teacher may want to ask parents to send in an envelope with cash so the teacher can set up a bank and perhaps take the students shopping in the community.

INTRODUCING THE CONCEPT

OBJECTIVE: The student will be able to simulate recording money earned (or received) and spent during a week or more.

MATERIALS:
- Recording sheet from Appendix B (Where Does the Money Go?)
- Transparency of recording sheet (for teachers of small groups—optional)
- Fine line marker (optional)
- 6 to 10 real receipts
- Pencils and calculator
- Realistic play currency

SUCCESS STEP: Ask the students to tell you the last thing that they bought at the store.

PROCEDURE:
1. Demonstrate what a receipt looks like by pointing out its parts. Explain that receipts are important because they help us keep track of what we have spent. (They are also necessary if we need to return things.)
2. Give the students some real receipts to look at and have them identify the parts—date, prices for individual items, the subtotal, the tax, and the final total.
3. Using 3 to 5 receipts, record the items on the recording sheet so the students can see. Either mark the prices on the overhead transparency or on the actual recording sheet.
4. Have the students record the same receipts on the Pay column on their recording sheets.
5. Then have the students add the receipts with their calculators.
6. Check their answers.
7. Repeat the process with another set of receipts.
8. Now have the students pay attention to the line on the recording sheet that says "Earn." Pay the students various amounts with the play currency. Have them count up the money and record it in the Earn column.

9. Point out the "Balance" column. Tell them that balance is a word for how much money they have left. Show them how to subtract the total amount paid out from the amount earned to get the balance left after spending some money.
10. Repeat the above exercise with more real receipts, if needed.

Note: The recording sheet is set up somewhat similar to a checkbook. One column is labeled PAY for the costs of items bought, while another column is labeled EARN for the deposits. The labels reflect back to the game Earn and Pay (see Chapter 4).

RECORDING ACTUAL EXPENDITURES (FOR OLDER STUDENTS)

It is important for students to have some reasons to record their actual spending. Teachers should enlist the parent's help with keeping receipts and recording the student's spending for the following activity.

OBJECTIVE: The student will be able to record money that he has spent over the period of a week and figure out his balance

MATERIALS:
- Receipts from a week's worth of purchases for each student
- Where Does the Money Go? (Appendix B)
- Pencil

SUCCESS STEP: Ask the students to tell you the most expensive thing that they bought that week.

PROCEDURE:
1. Have the student keep a record of a week's real spending by writing down each purchase or keeping receipts. Teachers may need to send a letter home to the family or supported living provider to help the student keep track of spending.
2. The student also needs to count the money that he has at the beginning of the week and record it under "Balance" on the Where Does the Money Go? form.
3. At the end of the week, ask the students to write each purchase on the Where Does the Money Go? form.
4. Have the students total the amount they spent, using a calculator.
5. Tell the students to subtract the total spent from the beginning balance.
6. You will have to find out if the students have any other sources of money besides their allowance or wages because these earnings must go in the **Earn** column. Help the students to understand that earning will make the balance get bigger. They must add earnings to the balance.
7. Continue recording weekly spending until the student can do this without direction.

An excellent book written directly for students with Down syndrome and other cognitive disabilities is *MY MONEY: Book One: Keeping Records,* available from Special Reads for Special Needs. The book provides many useful forms with simple instructions, introducing new skills in a very slow, deliberate manner. (See Resources.)

PLANNING SPENDING

OBJECTIVE: The student will be able to plan and record discretionary spending.

MATERIALS:
- Two Budget Planning sheets from Appendix B for each student
- Transparency of a Budget Planning sheet (if you are working with several students)
- Where Does the Money Go? from Appendix B
- Pencil
- Chalk or whiteboard

SUCCESS STEP: Ask the students to tell you something they bought last week. See if they can tell you what the item cost.

PROCEDURE:
1. Tell the students the following story, writing the numbers on the board.

    Nathan's allowance was $10 per week. He planned to buy a toy for his dog on Saturday for about $5. During the week he bought:
    - *one ice cream cone for $1.00*
    - *school lunches that totaled $2.00*
    - *one bottle of water for $1.00*

    On Saturday, he wanted to go to the pet store to buy the $5.00 toy for his dog. Did he have enough money to buy it?

2. Write the $5.00 pet toy on the planning half of the Budget Planning sheet Then fill out the bottom half of the sheet with the things that he actually bought. Total the spending and subtract the amount from Nathan's allowance of $10.00. Ask the students if there was enough allowance left to pay for the dog toy.
3. Have the students fill in their own Budget Planning sheet with Nathan's figures.
4. Have the students do the following case study on their own. Read the case study to the students and write the important numbers on the slate or whiteboard. Watch the students while they do the work and help them whenever they need it:

    Steve gets an allowance of $25.00 per week, but he needs to use some of it to buy his lunches. Steve wanted to go to the movies on the weekend. It would cost about $6.00. He needed $8.00 for lunches. He put these expenses on his budget-planning sheet and added to find the total.

    Steve bought the lunches he planned to that week. He also bought a keychain with a light for $3.00.

    Put all the items on the Budget Planning Sheet and see if he has enough money from his allowance to go to the movies. Follow these steps:
    - *Put the weekly allowance in the MY MONEY column on the Budget Planning Sheet.*
    - *Write down what Steve actually spent on the bottom half of the sheet.*

- *Then add up the money Steve actually spent.*
- *Subtract that amount from the total in MY MONEY.*
- *Does he have at least $6 left to go to the movies?*

5. The important part of this exercise is to emphasize the need to plan ahead of time for purchases. If the students are secure with the case study budget planning, you can start them planning with their own allowance or money earned. If not, make up a few more case studies for them to practice on.

INTRODUCING REAL BUDGETING

For students to begin generalizing budgeting to real life, it is important that they have some money to plan with. If it is not possible for the student to have actual money at school, set up a simulated situation with realistic play money. For example, pay students an allowance of play money for coming to school each day. Give them play money for completing assignments. Have some small items they can buy with the play money during specified times of the week.

OBJECTIVE: Students will record their earnings, plan their weekly expenses, and record their actual spending.

MATERIALS
- Where Does the Money Go? Sheet (Appendix B)
- Pencil
- Calculator

PROCEDURE
1. Continue having the students plan their expenses each week and then record their actual spending. Get family, roommates, and other people to remind the students to keep receipts, and stick to their planned expenses whenever possible. I have the students keep an individual three-ring notebook for their recording sheets with a pocket for receipts.
2. The amount of time spent on recording the students' spending will depend on their age and skills. I usually have the students keep a spending record for at least a month while I am teaching them other math skills.

Planning a Budget for Living

When a student can plan discretionary spending of a few items, you may want to begin teaching him how to budget for all his expenses—either for now or for the future. Many students do not have any idea of what it costs to live. For example, they have no idea that they will have to pay for where they live. The idea that one has to make mortgage payments or pay rent is foreign to them because they never see the money that goes to pay for a home or for the utilities that are needed. They see the money spent on food, entertainment, and clothes, but parents seldom involve their

children when they are paying for many of the other living expenses.

Even when a student starts earning money from a part- or full-time job, he often can spend most of his money on his own wants and desires, if he is still living in his parents' home. However, when the student moves into his own apartment or home, his earnings must go for all those living expenses with very little left over for discretionary spending. I use the game *Get a Life* (see below) to forecast what expenses can be and the decisions that have to be made.

The student also needs to understand the difference between needs and wants in order to make wise choices. This will probably take many discussions about what a person needs in order to survive and what he wants or desires. In this chapter, we will concentrate on the skills needed for money management, but the skills are not useful if the individual spends money as soon as he receives it with little planning for needs.

WANTS VERSUS NEEDS

OBJECTIVE: The student will be able to recognize the difference between wants and needs for managing his money.

MATERIALS:
- Slate, whiteboard, or chalkboard
- Chalk or markers
- Wants or Needs Worksheet (Appendix B)

SUCCESS STEP: Ask the students to tell you the basic needs for a person to live independently. Prompt them if they leave out something important. Ask them if they have any idea of what the monthly costs for those needs are. Let them guess what costs would be, but do not tell them whether they are close to actual costs.

PROCEDURE:
1. Tell the students a story about someone who has difficulties managing his money. You may use this story about Paul, below, to start, but later it would be better to use an actual situation or modify the study to fit your local community.

Paul was very excited when he got his first paycheck. He took his friends out for pizza. He bought a cool jacket that he had been looking at in the store window for several weeks. He went to the movies three nights in a row. By the end of two weeks, he was broke. He didn't have enough money to pay for the bus to go to work or to buy his lunches at work. He didn't have enough money to pay for his share of the cable bill. "I must have lost some of my money," he said. "I couldn't have spent it all."

2. Discuss how Paul spent his money. Did he spend it on needs? Emphasize that he did not know how his money had been spent. Try to help the students realize that if he had kept the kind of record they have been keeping, at least he would know where his money had gone.

3. Retell or reread Paul's story and have the students fill out the bottom half of the Money Budget sheet. Have them estimate what the costs of pizza for friends, three movies, and a jacket might have been. (You might be interested to find out how realistic their ideas of costs are.) Tell them what reasonable costs would be, if needed. Add up the cost of the items.

4. Discuss how Paul needs to set up a budget—that is, a plan for spending his money to make sure he has enough money for his real needs. If your students are young or do not have a job or way to earn money, but have an allowance, talk about using a budget to help keep them from wasting their money.

   If the students do not get an allowance, teachers can set up a token economy as described above where the students earn and pay out play money. If money is not really a difficulty with the family, it may be that parents just take care of all the student's needs and do not feel he is capable of handling money. Parents: no matter how young or inexperienced with money your child is, *I strongly recommend giving him an allowance.* All children need experiences paying for wanted items; this is especially true for children who need more repetition to learn.

5. Tell the students that Paul lived at home, but he had promised to pay part of his expenses to his parents. He pays part of the cable TV bill and part of the grocery bill for food. He has to pay for all of his entertainment (like movies), his lunches at work, and the daily bus fare needed to go to work. Paul makes $250 per month after taxes.

6. Write Paul's **needs** on the chalk- or whiteboard.
   Bus fare per month................$35
   Food per month....................$80
   Cable TV bill (part). $15
   Lunches ($4 per day)............$80   Total = $210

7. Have the students total Paul's needs for a month. Compare Paul's needs with his income from working. Paul has only $40 left after he takes care of the things he really needs.

8. List Paul's wants or discretionary spending. Have the students total Paul's discretionary wants.
   Entertainment.........................$40
   Eating out.............................$40
   Extra clothes.........................$40   Total = $120

9. Ask the students if Paul has enough money to spend $120 on "wants" or discretionary items. (No)

10. Have the students decide how Paul can cut his spending so his income matches his spending. For example, point out that if he packs a lunch some days and spends $50 on lunches instead of $80, he would save $30.

11. Point out that even if Paul cuts his spending, he will not save anything for future needs or wants. For example, Paul may want to get a cell phone that costs $100. He needs to save to buy that. (He also needs to consider that he will have more expenses when he has to pay a monthly cell phone bill.)

12. Have the students do the Wants versus Needs worksheet from Appendix B. There may be variations on what the students think are needs or wants. Have the student explain why that item is a need and you make a judgment on it.

BUDGETING FOR LIVING

OBJECTIVE: The student will be able to recognize or express the financial needs of a person living independently.

MATERIALS:
- Receipts or copies of checks used to pay for basic living expenses such as food, clothing, or utilities
- Chalkboard or whiteboard and chalk or marker
- Calculator

SUCCESS STEP: Have students look at the receipts and guess or read what they are for. Praise participation.

PROCEDURE:
1. Write a simple monthly budget on the board for an adult living on his own. Use the categories of:
    - Rent   _____
    - Food   _____
    - Utilities   _____
    - Clothing   _____
    - Bus fare   _____
    - Phone   _____
    - Savings   _____
    - Entertainment   _____

2. Discuss each category and explain why the costs are what they are. You may want to talk about a possible range for each cost. For instance, rent might cost from $300 to $650 (use costs in your area) depending on whether they have a roommate, etc. Find out about possible supported living options in your area. For example, my son pays one third of his income, no matter what that is.

3. Use a calculator to find the total monthly budget.

4. If you have students who are going to be transitioning from school soon, you may want to bring in a young adult who handles his finances while living in the community to talk to the students.

GET A LIFE GAME (SIMULATION OF LIVING INDEPENDENTLY)

OBJECTIVE: The student will make wise choices while simulating living independently.

MATERIALS:
- Get a Life game from Appendix B
- Game tokens
- Dice
- Currency (play)
- Post-It Notes

SUCCESS STEP: Have the students read the labels on the various squares on the Get a Life game board. Show them that the squares inside the game track are optional "wants" such as pizza that they can buy when they land on the square.

PROCEDURE:
1. Give the students $250 for their wages for two weeks of work. (Most of the jobs that high school students and young adults get pay them every two weeks.) Explain to the students that we have talked about budgeting in costs per month. However, we are going to pay our bills two times a month because we get paid two times a month. For example, we pay rent two times on the game board.
2. The students toss a die and move their game tokens as indicated.
3. The students have to pay the rent, food, and utilities whether they land on those squares or not.
4. They pay or receive money according to what is written on the squares they land on.
5. Whenever they move next to a discretionary item such as "Buy a CD," they can choose to buy it or not. If they buy something, the teacher gives them a Post-It Note with the name of the item.
6. The winner is the first person to get to the finish (must toss exact number with die).
7. However, the second winner is the one who has the most money left at the end. Some winning depends on the luck of the dice, but the students usually find that they have a better chance to be the second winner if they do not buy many of the fun items that they have an opportunity to buy.
8. The game gives students an opportunity to learn about needs and wants and living independently. This game can be used over and over to provide a simulation of what it is to live independently—and they will enjoy it too.

The MY MONEY series of books published by Special Reads for Special Needs has a book on *Keeping a Budget* (Hale, 2004). This series is written directly to the student with a cognitive disability and has simple, clear instructions that progress slowly.

Banking Basics

Middle and high school students should learn about savings and checking accounts, and, if possible, open their own accounts at a bank. Parents may help their son or daughter open an account at a local bank. Schoolteachers may help their students to open an account at a bank close to the school. Since the accounts of school students will probably be very small, look around for a bank that has free student accounts with no minimum balance or is willing to waive any fees.

Banking requires a new vocabulary to be understood: deposit, checking account, interest, balance, savings account, endorse, withdrawal, bank statement, and ATM. You can explain what the words mean as you discuss banking and handling your money. If the student has learned to record his earnings and what he spends in the previous section called "Where Does the Money Go?" he will find that recording amounts in his check register is quite similar.

Some families want to get their children in the habit of saving and open a savings account for them when they are quite young. Children can learn about making a deposit and earning interest when they are accumulating savings. However, using a checking account is a more useful skill and more rewarding to the student than making regular deposits in a saving account.

DEPOSITING MONEY—CASH ONLY

OBJECTIVE: The student will be able to fill out an accurate banking deposit slip for cash only.

MATERIALS:
- Banking deposit slip (from your bank or one from Appendix B)
- Play money

SUCCESS STEP: Ask the student to pick $100.00 from the play money.

PROCEDURE:
1. Explain why people put their money in a bank. For example, so it won't get stolen, so they can earn interest (money the bank gives you for the use of your money), so they won't spend money they need to save, or other reasons.
2. Tell the student that he is going to deposit $100.00 in the simulated bank. Explain that a deposit slip is needed to put money in the bank and that the teller will give them a receipt to show that they have put this money in the bank.
3. It will be more fun if you pretend to be a teller and sit behind a desk.
4. Pull out a banking deposit slip and have the student fill in the necessary spaces.
5. Write in the date.
6. Write in the amount that you are going to deposit next to the word currency ($100.00). Explain that currency on a deposit slip means paper money and coins, but not checks.
7. Write the same amount ($100.00) in the space marked *Deposit Total*.
8. Have the student give you the slip and his money.

9. Pretend to be the teller and give the student a receipt. (You can photocopy the deposit slips onto a different color of paper to use as receipts.)
10. Repeat the procedure several times with different amounts of currency. See if the student can fill out the deposit slip independently.
11. Using a simple withdrawal slip, explain (demonstrate) how you can get your money out of the bank after you have deposited it.

DEPOSITING A CHECK AND WITHDRAWING CASH

OBJECTIVE: The student will be able to simulate depositing a check in a bank account and withdrawing cash.

MATERIALS:
- Simulated payroll check for $344.42 (Appendix B)
- Deposit slip (Appendix B)
- Play money

SUCCESS STEP: Ask the student to point out the parts of a deposit slip from Appendix B.

PROCEDURE:
1. If necessary, explain to the students what a check is. (A piece of paper that enables you to get money from somebody else's bank account if that person signs the check.)
2. Tell the student that you are going to give him a play payroll check (explain what a payroll check is, if necessary). Explain why a person might want to deposit part of a check, but also take out some of the money in cash. Tell him that when he deposits the check, he can take $10.00 out to spend.
3. Have the student endorse the check on the back. Explain to him what "endorse" means.
4. Help the student fill out the deposit slip.
- Write in the date.
- Write in the amount of the check in the space provided ($344.42).
- Write the amount of money you want to take out ($10.00) in the space marked *Less Cash*. (Sign name or initials if space is provided.)
- Subtract the cash received from the total amount to be deposited and put the answer in the space for *Deposit Total*.
5. Pretend to be the teller and give $10 to the student and a receipt for the amount of the deposit ($334.42).
6. Repeat the steps above using different check and currency amounts until the student is able to do the steps independently.

LEARNING TO USE A CHECKING ACCOUNT

OBJECTIVE: The student will be able to fill out a check accurately and describe the purpose of a check.

MATERIALS:
- Check forms (Appendix B)
- Transparency of check form or large check from Appendix B
- Worksheet on Writing Out Check Numbers in Words from Appendix B

SUCCESS STEP: Ask the student to write his name and address and phone number on a piece of paper like he would like to have on his checks. (If he is unable to write, have an adult write the label as he dictates it.) Praise completion. (If possible, make labels with this information on the computer so the student can have personalized checks.)

PROCEDURE:
1. Show the large labeled check or the transparency to the students.
2. Point out the following items:
    - Name and address of person writing the check
    - Date
    - Check number
    - Who will get the money from the check
    - Amount of money in numbers
    - Amount of money in words
    - Name and address of bank
    - Memo—what the check is for
    - Signature
    - Bank numbers

    If necessary, provide another check as an example and have the students point to the parts as you name them.
3. Perhaps the most difficult part of writing a check is writing the amount of the check in words. It is probably the only time we ever have to write out large numbers in words. See if the student is able to write dollars and cents in words. If the student has difficulty, it might help to review the Place Value Chart on Money from *Book 1:*

HUNDREDS	TENS	ONES	DECIMAL POINT WALL	CENTS	
Grown-ups	*Teens*	*Kids*			
Two Hundred	thirty	five	AND	sixty	six
2	3	5	•	6	6
			•		

The *Kids* get allowances in one-dollar bills. Explain that the *Kids* are closest to home, which is shown by the wall (which is a wall representing the decimal point). The *tens* column is for *Teenagers* who get allowances in ten-dollar bills. The *hundreds* column is for *Grown-ups* who get their money in hundred-dollar bills. We read the number by starting on the left with *Grown-ups*. *Grown-ups* are kind of snotty; you have to say their last names (hundred). Next you read the *Teenagers and Kids* as **a two-digit number.** The arrow on the chart shows that the two numerals are read together as a two-digit number. Then you are finished with the dollars and go on to the cents. You say *and,* and read the cents as **a two-digit number.** The number above is read, "Two hundred, thirty-five dollars and sixty-six cents." Have the students practice writing out dollars and cents amounts.

4. Have the students do the worksheet on Writing Check Amounts in Words. The first part has the numbers written on the worksheet. The second part has blanks for you to fill in for numbers that may be important for the student.

5. If the student is just not able to write the check amount in words, he may be able to write the number again in numbers on the line where the words are normally written. Some banks in our area would take the check—if the two numbers were exactly the same. You might ask the bankers in your area if this would be possible.

6. Make up different scenarios and have the students practice writing in various amounts of money to various people. For example, you want to give your nephew a check for twenty dollars for his birthday. Or, you want to pay your neighbor ten dollars for feeding your cat.

It is important that the students understand that writing a check is not a way for them to get new money. Ask them if it would be wise for them to carry a lot of money in cash in their pockets or purses every day. Most of them will realize that someone could take their money or they could lose it. Tell them that if they put their money in the bank and wrote a check to take some of their own money out, their money would be safer. It is also a safer way to send money through the mail, such as when you are paying a bill or ordering something from a catalog. If the students are young and do not have ways to earn money other than their allowance, they may have little use for this information. Older students, however, may see the need for banks and checks, and they probably will have more experience with their parents writing checks. Emphasize that they can only write checks for the amount of money that they have put in the bank. (Don't mix them up by telling them about overdraft protection.)

The MY MONEY series of books also has a student book on *Keeping a Checking Account* (Hale, 2005) which uses a check register with bigger spaces than a regular checking register and illustrates rules for banking.

It takes a good deal of practice to be able to write checks and record them accurately in a check register. However, the most important part of using a bank and checks has to do with a person's attitude and responsibility. Parents are the ones who have to judge whether their son or daughter is ready to use a checking account.

Debit Cards

When I saw my son's eyes the first time he watched me getting cash at the ATM machine, I thought I would never want him to use a debit card. He said, "Money is coming out of the machine!" I could see that he thought that the machine was making money—free. I tried to explain—rather ineffectively.

I expressed my fears of introducing my son to ATM machines at several presentations on teaching math that I made to parent groups. Several parents explained to me that they had been very successful in teaching their son or daughter to use the ATM. One advantage they mentioned was that the ATM machine gave out a receipt showing how much money was still in the checking account, even if their child couldn't or wouldn't keep track of the balance in his account by using a check register. Another advantage they cited was that their child couldn't get money from the ATM if there wasn't money in the account. (This is the case only if you have a debit card that cannot also be used as a credit card. These combination debit/credit cards require more education to use and can be too much of a temptation to spend more money than you have.) One family has their son use a debit card primarily for purchasing medications. Debit cards might also be handy if your adult child cannot get to a bank easily but there are ATM machines in nearby grocery stores, etc. I can handle the idea of that kind of debit card and my son can also.

It may be somewhat difficult to teach students with Down syndrome or other cognitive disabilities the skills involved in using an ATM. Every bank seems to have a slightly different system. Parents can take the student with them when they make a withdrawal or deposit and show the student what they are doing. There are now some play ATMs (see Resources section) on the market that could be used if the student would be using the ATM frequently. The toy ATM can be used as a bank also.

Credit

Using credit responsibly is a problem in our society for many people—with and without disabilities. Research has shown that people tend to spend more when using a plastic credit card than when using cash to pay for things. Many parents and educators feel that people with moderate to severe cognitive disabilities should not use credit because of the temptation to spend more and the possibility that others may take advantage of them. However, it is hard to avoid using credit when buying large-ticket items such as major appliances, furniture, and computers or TVs. It may be necessary for your adult child to establish a credit rating so he can live independently or in a supported setting. Some parents might also want their children to have a credit card to use in case of an emergency—for instance, if their adult child is traveling out of state alone or enrolled in some kind of postsecondary school away from home.

A student can learn about credit more easily at home than in a school setting. Usually the student has seen his parents use a credit card for shopping. The parents can explain that the family has to pay for the item when the bill comes every month. They should show the credit bill to the student at home when it arrives and explain how they are going to pay the bill. They can explain about spreading the cost of the item over several months and why the credit is useful.

Parents must also talk to their child about the interest that must be paid if you spread the cost over several months. I sometimes call interest the rent that you have to pay for using someone else's money (if the student understands the concept of rent). Parents can also explain the importance of not getting into heavy credit card debt. It is important that students learn the consequences of credit card debt.

ROLE PLAYING CREDIT CARD USAGE

This activity provides a simulation of using a credit card. When doing the activity, it is important to use situations that will sound real to your student(s). Use props and *ham* up the situation a little. Change the story to suit your situation.

The adult can do a role-play of this example and have the students help, either representing the credit card company or Kayli. Be sure to have a chalk- or whiteboard available to write down the numbers. Have the students follow along using their calculators, if possible. Try to have the students empathize with Kayli, and see what can happen with credit card debt.

OBJECTIVE: Students will be able to explain some of the consequences of using a credit card improperly.

MATERIALS:
- Calculator
- Chalkboard or Whiteboard
- Chalk or markers

PROCEDURE:
1. Read the case story below aloud, writing key amounts on the board:

Kayli had been saving for a jacket that cost $80, but she only had $30 saved. Then she got a credit card in the mail. She got excited! She thought she could buy the jacket now. She bought the jacket with her credit card the next day.

A couple weeks later, Kayli got a bill from the credit card company for $80. She still did not have $80 saved. The bill said the minimum payment was only $10. She knew that the minimum payment was the smallest amount that the credit card company would accept. She sent in $10. She thought, "Now I have only $70 left to pay." However, when the next bill came in a month, the bill was for $82.60. The extra $12.60 was the interest that Kayli had to pay because she still owed the credit card company $70. If Kayli had paid off the entire $80 when she got the bill in February, she would not have had to pay the extra $12.60. She could have been able to use her credit card safely without a cost to her.

Kayli thought, "My jacket cost $80, but with the interest I have to pay, it really cost me $92.60. (The original $80 plus $12.60 interest equals $92.60.) Was it worth getting the jacket some time earlier, but having to pay $12.60 more? Next month I will have to pay more interest to the credit card company if I still owe them money."

2. Ask the students:
   1. What should Kayli do now? (Pay off the bill as fast as possible.)
   2. If Kayli only sends in the $10 minimum payment again, will she have the bill paid off?

No, her bill is now $82.60. If she only pays $10, she will still owe $72.50 plus the interest for that month ($13.07) for a total of $85.57. She will still owe more than she paid for the jacket, even though she has paid $20.

3. Continue with the story:

Kayli got sick around the first of March right after she got her bill. She forgot to pay her credit card bill. In April, her bill was for $85.60 plus interest of $15.41, plus a late fee of $12.50 for a total of $113.51. Banks usually charge late fees if the payments are not in on time. Now she owes $113.51 for a $80 purchase.

One reason Kayli owes so much money is because her credit card company charges her quite a lot of interest every month (18% or $15.41) for her to use their money. Kayli may be able to get a credit card that charges less interest, such as 10-12% ($8.60–$10.27) interest, if she looks at other companies. However, she may need help from family or friends to find those companies. (We will get into computing percentages in Chapter 17.)

4. Ask the students: How can Kayli get out of this debt? Brainstorm some ideas. Then:
   Tell the students that when they use credit, companies called credit bureaus keep track of what they spend, if they pay on time, and if they have too much credit card debt. This is called a credit report. When you are trying to buy some expensive things such as a large TV or when you want to rent a house or an apartment, stores or property owners can look at that credit report and see if they think you are a good credit risk—if they think that you will pay back the money promptly. They can refuse to give you credit, especially if you have not paid your bills on time. Sometimes you can get credit if a parent or responsible person will co-sign the credit application. That person promises to pay the credit card company if the student can't pay.

5. Ask the students what happens if your credit card gets stolen. See if they understand that they need to call the credit company immediately. Tell them that they will not lose as much money if they call the company right when it happens. (You can explain the situation in your area about limits that the credit card owner must pay.)

6. Have the students tell you two reasons why a person their age should have a credit card and two reasons why they should not have a credit card.

CHAPTER 16

Geometry and Data

Questions to be answered:

Can the student:

1. Identify the shapes of circle, square, triangle, and rectangle.

2. Point to or identify a ball (sphere), a box (cube), and a tube (cylinder).

3. Identify intersecting lines, right angles, parallel lines, perpendicular lines (parallelograms, rhombus, trapezoid–if you want to impress someone).

4. Identify above types of lines on a real map of your area.

5. *Optional:* Define acute and obtuse angles.

6. *Optional:* Demonstrate use of a protractor for measuring simple angles.

7. *Optional:* Define a triangle and its interior angles.

8. Show the circumference, diameter, and radius of a circle.

9. Keep data on her own hours of TV watching for a week.

10. Make a line and a bar graph of the data.

11. Count different types of food or candy and graph the resulting data.

Teaching Math, Book 1, includes a chapter on identifying two-dimensional figures such as circles and triangles and some three-dimensional figures such as cubes and cylinders. If your student does not understand those basic concepts, you should refer back to Chapter 19, Shapes and Patterns, in *Book 1.*

Usually a student who continues with geometry does much work with measuring areas, perimeters, and degrees of different angles. Measuring areas and perimeters is covered in Chapter 12 of this volume. Measuring angles does not seem to have much "survival" value to students who are concrete learners. However, if a student is included in regular math classes, she will likely need to learn at least some of the basic concepts related to angles in order to access the regular curriculum.

We begin with some review of the geometry covered in *Teaching Math, Book 1.* We next get into some more advanced geometric concepts that you might want to use to help a student access the regular math curriculum.

REVIEW OF SHAPES—*THE THINKING GAME*

OBJECTIVE: The student will be able to identify the shapes of circle, square, triangle, and rectangle.

MATERIALS:
- A set of cardboard shapes (circle, square, triangle, rectangle) for each student (from model in Appendix B)

SUCCESS STEP: Have the student hold up the matching shape when the teacher holds up the model.

PROCEDURE:
1. Have the students lay their cardboard shapes in front of them on the table or desk.
2. The teacher should say, "Hold up the circle" (without holding up a model). The students should all hold up their circles. Repeat with the other shapes.
3. Then, in a mysterious voice, say, "I am thinking of a shape. It has four sides. The sides are the same length. Do you know what I am thinking?"
4. The students should hold up the appropriate shape. Praise them saying, "You are great thinkers." Later, ask each student separately to tell you the name of the shape you are thinking of.
5. Repeat in random order for the other shapes. Recognizing the attributes of a shape from a verbal explanation is much more difficult than just recognizing shapes visually. The student must make a mental picture from the teacher's description. If they are unable to use the clues to recognize the shape, go back to steps 1 and 2.
6. If the students are able to identify the shapes, go on to the three-dimensional objects below. If not, review shapes.

THREE-DIMENSIONAL SHAPES—*INTRODUCING THE CONCEPT*

OBJECTIVE: The student will be able to identify a ball (sphere), a box (cube or prism), and a tube (cylinder).

MATERIALS:
- One ball
- One cardboard tube
- One box (taped closed)

- Whiteboard and marker (or slate and chalk)
- A large box filled with packing bubbles or newspaper to hide the shapes

SUCCESS STEP: Have the students feel each of the objects. Ask them if they can see some similarities to the circle, square, or rectangle. They should be able to relate the ball and tube to the circle and the box to a rectangle or square. If they are unable to do so, do the following procedure, but talk to them about the similarities as you proceed.

PROCEDURE:
1. Hold up each object and tell the students the proper geometric name for it.
2. Print the name on a slate or whiteboard and show the object again.
3. Repeat step 1 and 2 until the students can identify the objects.

DIG IT GAME

1. Hide a ball and another of the three-dimensional shapes in a large box and fill it with packing peanuts, tissue paper, newspaper, or something else that will hide the shapes from view.
2. Blindfold the student.
3. Have the student dig for the ball. (Call it a sphere, if she is working on that vocabulary.) Help her to feel the roundness of the ball if she has trouble.
4. Repeat with digging for the boxes (prisms) or tubes (cylinders).
5. This is not as easy a task as it seems. The student has to picture the shape in her mind and distinguish it from the others and then feel for it with only her fingers.

Lines

Having a basic understanding of vocabulary related to lines can be a useful survival skill for students who travel from place to place alone and must sometimes ask for directions or interpret a map—for instance, at a mall or an amusement park. Of course, if they

are included in general education math classes, they will also need to learn about different kinds of lines in geometry.

Math Vocabulary

If your student is adept at math or is included in the general classroom for math, you may want to teach her more Geometry vocabulary words. You can teach the names and definitions with flashcards, by matching explanations with drawings, or by using mnemonic clues. Here are some hints:

- An **angle** is formed by two lines (rays) meeting at the same end point. An angle is measured in degrees with the symbol of °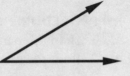

- **Right angles** are like square corners (L shaped).

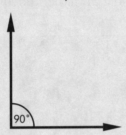

- **Parallel lines** are always the same distance apart. They never meet.

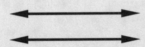

Ask the students to model their hands (palms facing each other) and arms going parallel no matter where they move.

One way that might may help the student to remember the vocabulary word would be to show that the word **parallel** has two parallel lines in the middle of the word—the two l's.

- **Intersecting Lines:** Lines that meet each other and continue on. The students may have heard the term **intersection** in connection with roads.

- **Perpendicular:** Two lines meeting at a right angle to each other. They form a T or a plus.

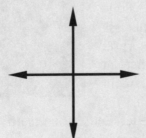

LINES—*INTRODUCING THE CONCEPT*

OBJECTIVE: The student will be able to show, with concrete materials, intersecting, parallel, and perpendicular lines.

MATERIALS:
- Two rulers, Wikki sticks, pipe cleaners, pieces of licorice (Twizzlers), pretzel sticks, or other long, slender objects for each student
- Whiteboard and marker or chalkboard and chalk

SUCCESS STEP: Draw a line on the board and ask the students what it is. Praise correct answer.

PROCEDURE:
1. Tell the students that you are going to be learning about some special kinds of lines. For today, you are going to use rulers (licorice, pipe cleaners, etc.) to represent lines.
2. Hold up 2 rulers so they are intersecting in an X in front of you. Ask the students to do the same with their rulers. Ask them what they have made. If they say "an X," agree with them. Then tell them that in geometry, when 2 lines meet, you say they are "intersecting."
3. As you continue to hold the rulers together in an X, slowly begin to swivel one of the rulers around so you change the angle between them, but so they are still touching at the midpoint. Ask the students, "Are the lines still intersecting?" (Yes). Move the rulers all around, keeping them intersecting, while you ask the students if the lines are still intersecting. Have them do the same with their own rulers.
4. Separate the rulers and hold them so they are no longer touching. Ask the students if they are intersecting. (No).
5. Put the rulers back together so they form a plus shape. Ask the students to copy your shape with their rulers. Ask them what they have made. If they say "a plus," or "a cross," agree with them. Also agree with them if they say you have made intersecting lines. Tell them that these are special intersecting lines. When lines intersect in a plus or cross, they are called "perpendicular lines." Show them the sharp, square corners where the rulers are touching.
6. Finally (not necessarily on the same day), hold your rulers up so they are parallel to each other. Ask the students to do the same. Ask them if these lines are intersecting. (No). Tell them that when lines are side by side and do not intersect, we call them "parallel lines" (like parallel bars, if they are familiar with gymnastics equipment).
7. At the end of the lesson, if appropriate, hand out pretzel rods or licorice sticks. Ask the students to make the various lines with pretzel rods or licorice sticks on a paper plate. Then let them eat the licorice or pretzels.

LINES ON A MAP

OBJECTIVE: The student will be able to determine whether roads on a map or actual roads intersect, and/or are parallel or perpendicular.

MATERIALS:
- A simple map of your area
- Definitions of parallel, perpendicular, and intersecting lines with examples displayed where the students can see them (see box above)
- A black or red marking pen
- A highlighting pen

SUCCESS STEP: Have the students name the streets that run near their house or school. (If they can't name any, try to find time to take them on a walking field trip and point out the names of the important streets.)

PROCEDURE:
1. A real map of your area will probably have too much detail on it for the students to be able to see the important lines. Using a marker, draw on the map the major streets that you want them to see. Label the important streets with their names. You may want to do this with the students watching and have them identify the streets as you draw them.
2. Using another color marker or highlighter, circle the lines that you want them to label (perpendicular, parallel, etc.). Remember that some of the streets can fit into more than one category (perpendicular lines may also intersect).
3. Keep the map hanging up for a few days. Teach the students the names and categories of the lines representing the streets.
4. If possible, take the students on a walking field trip where they can identify the real streets that the lines in your map represent. Point out where they intersect or where one is perpendicular to another.

Other Math Vocabulary (Optional)

If your student is adept at math or is included in the general classroom for math, you may want to teach her more Geometry vocabulary words. You can teach the names and definitions with flashcards, by matching explanations with drawings, or by using mnemonic clues. Here are some hints:

- A *parallelogram* has four sides with two pairs of opposite, parallel sides.

- A *rectangle* is a parallelogram with 4 square (right angle or L) corners.

- A *rhombus* is a parallelogram with all sides equal in length, but no right angles.

- A *trapezoid* is a four-sided shape with only two sides parallel.

Angles

A lot of attention in geometry and later on in trigonometry is spent on defining and measuring various angles and their relationships to each other. For students, it can be exciting to measure an angle and determine its relationship to another angle. If the students are getting math instruction in the general education classroom, you may need to use hands-on methods to illustrate the meaning of the information about angles. You may also want to give the students some hints to help remember the words and their meanings (see below).

For survival math, students really only need some basic information about angles, so use your discretion in deciding how much, if any, of the information below you require them to learn.

Angle Vocabulary Hints

If your student is adept at math or is included in the general classroom for math, you may want to teach her more Geometry vocabulary words. You can teach the names and definitions with flashcards, by matching explanations with drawings, or by using mnemonic clues. Here are some hints:

- An *angle* is formed when two lines (rays) meet at the same end point (vertex). Angles are measured in degrees. A straight line is an angle of 180 degrees.

- *Acute angle*—an angle that is less than a right angle (less than 90 degrees). Perhaps it can be remembered as "a cute" little angle like a cute little girl (smaller than a woman, less than a right angle).

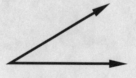

- *Obtuse angle*—an angle that is bigger than a right angle (more than 90 degrees). Can be remembered like an *obese* angle—one than is fatter (larger) than a right angle.

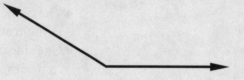

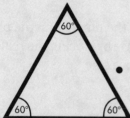

- *Triangle*—three-sided figure whose angles add up to 180 degrees.

- *Protractor*—a D-shaped instrument used to measure angles.

MEASURING ANGLES WITH A PROTRACTOR *(OPTIONAL)*

OBJECTIVE: The student will be able to measure the degrees of an angle with a protractor.

MATERIALS:
- A protractor for each student and teacher
- Measuring Angles worksheet from Appendix B

SUCCESS STEP: Have the student find and point to the two zeroes that are on the protractor. Have the student trace each scale from zero to 180. Show him how you always use the scale that starts with 0 to measure an angle, whether the angle opens to the left or right.

USING A PROTRACTOR:
- Note circle at middle.
- The point (vertex) of the angle is put under the circle at the middle of the straight side.
- One side of the angle is lined up with the straight line that is across the bottom of the protractor. That angle line should point to either the right or left side of the protractor and to the 0 mark on the scale. (Ignore the 180 that is also written there.)
- Look at the other side of the angle and line it up with the number on the protractor that it is closest to. You can tell what scale to use by tracing the scale that starts at the 0 mark that you identified in the above step.

PROCEDURE:
1. Demonstrate how to measure an angle by putting the point (vertex) of the angle exactly under the little open circle on the bottom edge of the protractor.
2. Line up one side of the angle so it is even with the line on the bottom of the protractor.

3. Identify the scale that will be used to measure the angle by touching the 0 near the end of that angle side.
4. Follow the line on the other side of the angle to the number that it is closest to on the scale you identified above.
5. Write down the answer using the degree sign for the students to see.
6. Have each student do one angle from the worksheet while you check that she is following the steps you have taught.
7. Practice using the angle worksheet until the students feel confident that they can measure angles this way. (Be sure to use right, acute, and obtuse angles for the practice.)

Circles

Whether students are learning survival or general education math, they will have to know some things about circles. I will list some of the more commonly taught ideas, but I will not require the student to draw circles with a compass. Students with fine motor difficulties often have trouble holding the point of the compass in the center and twirling the pencil about the sharp point with a twist of the wrist to draw the circle. In addition, the sharp point on the commonly used student compasses can be dangerous.

If you know your student will need to draw circles, and she has some fine motor difficulties, you can order a compass that you operate with two hands. These compasses generally consist of a flat piece of plastic somewhat like a ruler. You use a finger on one hand to hold it stationary at one end and insert a pencil point into a hole at the desired distance and then rotate it around in a circle with your other hand. The Triman safe drawing compass, Bullseye SAFE-T compass, and GeoTool Compass are possible options (available from ETA Cuisenaire at www.etacuisenaire.com, Learning Resources at www.learningresources.com, or EAI Education at eaieducation.com).

Drawing a circle with a Bullseye SAFE-T compass.

CIRCLES

OBJECTIVE: The student will be able to recognize the circumference, diameter, and radius of a circle.

MATERIALS:
- 2-3 small plastic, disposable plates (in 2 or more sizes) for each student
- Tape measure
- 1½ ft. of string

- Chalkboard, slate, or whiteboard for teacher
- Chalk or marker
- Straight ruler
- Paper and pencil for each student and teacher
- Calculators, if needed

SUCCESS STEP: Hand the student a paper plate and ask her what shape it is. Stress that a circle is a shape where all the lines are at an equal distance from the center. That is, every part of the outside line is at an equal distance from the center.

PROCEDURE:

1. Hold up a plate and say, "I want to know how many inches it is around this plate." Take the ruler and show them how much trouble it is to try and use the wooden or plastic ruler to measure something that is round.

2. Tell them that the distance around a circle is called a *circumference.* Ask if anyone can think of a better way to measure the *circumference* of that plate. They may tell you to use a tape measure or to measure a string that has been wrapped around the circumference of the plate. If they do not think of these ways, you demonstrate with the tape measure and then with a string.

3. Let each student measure her own plate using either a tape measure or a string and a ruler. Have them write down the number of inches that they measured.

4. Compare answers and see if the students agree. Re-measure if they don't.

5. Now use a marker to draw a straight line completely across a large plate and label it *diameter.* **Make sure the diameter runs through the center of the circle.** Hold up the plate and ask the students to pretend that this is a racetrack circle. Which would be faster—to run around the circle or to run straightly across the circle? (run across—the diameter)

6. The teacher should then measure the diameter and compare it to the circumference. The circumference should be a little more than 3 times the diameter.

7. Have the students measure the diameter of their circles. They can use their calculators and multiply the diameter by 3 and see if the answer is close to the amount of the circumference.

8. Have them do the same thing to their other circles. Write on the board: The circumference (distance around) of a circle is about 3 times the diameter (line straight across).

9. If possible, draw a circle on the playground and then draw a line for the diameter. Have the students "step off" around the circle and count the number of steps they take. Then have them "step off" the diameter of the circle. Write down the measurements and take them into the classroom. Again have the students multiply the diameter measurement by 3 and see if it is close to the circumference distance. (For younger students, you could also have one student race around the circumference while another races across the diameter to illustrate how much shorter the one path is than the other.)

10. Practice the above rule and the definitions a little bit each day so the students can remember them automatically.

11. If the students will have to know more detail about circles, you can introduce the term *radius* representing half of the diameter and the figure pi.

GENERALIZATION ACTIVITIES

- When showing the student how to walk somewhere, point out how two roads (or halls) are *parallel* to each other or *intersect* at a given place. Knowing which roads are parallel (go in the same direction) can be especially handy if the person gets lost or the bus takes a different route than usual, etc.
- Explain expressions that use geometry terms that are encountered in every-day life or on the news: "She did a 180."
- Have your student help you dig rows in your family's or the school's garden. Perhaps some of your rows will be parallel to each other, and others will be perpendicular.
- Play tic-tac-toe. Talk about how you have two sets of parallel lines that intersect at right angles to each other.
- When putting together a jigsaw puzzle, ask the student to help you find the pieces that make right angles (the corner pieces).
- Measure how much room you have for a new rug, table, or anything else square or rectangular and see how much area it can take up.
- When ordering pizza, talk about whether the sizes (12", 16", etc.) refer to the pizza's diameter or circumference. Likewise, when baking a cake or pie and the recipe calls for a pan of a certain dimension (e.g., 9") discuss whether that is diameter or circumference.

Data and Graphing

One of the contributions of the National Council of Teachers of Mathematics (NCTM) in setting up math standards for students in kindergarten through grade 12 was to look more closely at the importance of collecting data and graphing that data. We now understand that even very young students can collect and graph data to make it more meaningful to them. For people with Down syndrome or other visual learners, learning to make and read graphs can be very useful in helping them understand data in their daily lives, in the newspaper, etc.

TYPES OF DATA TO BE COLLECTED

In teaching students with Down syndrome or other hands-on learners about data, it is important to collect data that are meaningful to them or that are a vital part of learning in other academic areas such as science. In a classroom, students can make surveys of various items of interest such as:

- Physical comparisons of hair color, eye color, types of shoes, and clothing worn, etc.
- Preferences such as favorite TV shows, books, sports, colors, or foods.
- Circumstances such as birthdays, number in family, allowance amounts, number of soccer practices, and trips in the car per week.

They can also learn to use data to:

1. *Help them make decisions:* For example, when planning a party or other event, they can gather data on people's preferred beverages or movies and use that information to determine what drinks

to buy or what DVDs to rent. Or, when figuring out whether it makes more sense to take or buy their lunch every day, they can keep track of the costs involved for a month and see which alternative is more economical.

2. *Keep track of their progress toward meeting goals:* For example, if they are trying to increase the words per minute they can type or the number of steps they walk per day, they can chart that information and see whether they are making any improvement.

3. *Understand their academic progress:* Many school systems in the US are using *Dynamic Indicators of Basic Early Literacy Skills* (DIBELS) to measure early reading skills (Kaminski & Good, 1996). An important component of DIBELS is regular data collection with the student graphing the data, if possible. Another trend in student evaluation is Curriculum-Based Measurement (CBM), where the teacher frequently tests students on what she is teaching and adjusts her teaching quickly to address the students' needs. CBM also has both the teacher and student do data collecting and graphing.

Parents can collect data about their child's activities and behavior at home, such as:

- Household jobs done
- Times spent watching TV or on the Internet
- Times teeth brushed, bed made, clothes put in hamper, etc., without prompting

Of course, this type of data will not be very interesting to your child unless there is a reward involved. Graphing the data and posting your child's progress will most probably help her to stay interested in the tasks, or at least help her to remember what the goals are.

DATA COLLECTION AND GRAPHING

OBJECTIVE: The student will be able to keep data on her own hours of TV watching and graph the hours daily for a week.

MATERIALS:
- Paper and pencil
- TV
- Poster board and marker
- Data form (Appendix B)

SUCCESS STEP: Ask the student what TV shows she watches every day.

PROCEDURE:
1. If you are not the student's parents, ask the parents to help the student record the amount of TV that she watches every day for a week.

2. When the student has recorded 7 days of TV watching, her teacher or parent can help her make a poster showing the TV time. The graph needs to have the days of the week across the top of the graph and perhaps 30-minute intervals on the other axis, up to 8 hours. For example:

TV Watching

	Mon.	Tues.	Wed.	Thurs.	Friday	Sat.	Sunday
5 hr.							
4 ½ hr.							
4 hr.							
3 ½ hr.							
3 hr.							
2 ½ hr.			•				
2 hr.				•			•
1 ½ hr.	•						
1 hr.		•			•	•	
½ hr							

At first, the student may need some assistance in figuring out where to put the dot on the chart, and later she may need some supervision so the data will be graphed accurately. Connecting the dots will make a line graph that shows change over time.

Bar graph with same data

	Mon.	Tues.	Wed.	Thurs.	Friday	Sat.	Sunday
5 hr.							
4 ½ hr.							
4 hr.							
3 ½ hr.							
3 hr.							
2 ½ hr.			▓				
2 hr.			▓	▓	▓		▓
1 ½ hr.	▓		▓	▓	▓		▓
1 hr.	▓	▓	▓	▓	▓	▓	▓
½ hr	▓	▓	▓	▓	▓	▓	▓

3. You can make the charts at home or the student can bring in the raw data and make the charts under teacher supervision at school.

MORE DATA COLLECTION

For most students, an appealing reason to keep and graph data would be to have the opportunity to handle small pieces of food or candy. (Whether they get to taste any of the candy is up to you!) The following procedure was adapted from "A Taste for M&M's" on Ivars Peterson's Mathland web site (www.maa.org/mathland).

OBJECTIVE: The student will be able to count different types of food or candy and graph the resulting data.

MATERIALS:
- Snack food or candies that have varying shapes or colors such as:
  - 2 bags each M&M's (with various colors)—from Mars, Inc.
  - 2 rolls each Smarties (with various colors)—from Ce De Candy, Inc.
  - 2 boxes of animal crackers (with various animal shapes) from Nabisco Brands (Barnum's Animals)
  - Or another similar food that has various colors or shapes
- Paper and pencil

SUCCESS STEP: Put out a blanket or towel on the floor or use the table with a tablecloth. Have the student sort the candies or crackers into categories by color or by animal shape.

PROCEDURE:
1. Check to see that the snack is separated into categories correctly.
2. Have the students count each pile of food and write the count on a piece of paper (labeled with the color or shape).
3. Repeat the process with the second bag or box.
4. Have them put the data on a simple chart:

Color/shape	1st bag	2nd. bag	Total
brown	14	15	29

5. When the chart is finished, have them decide what sort of graph they want to make about it. Have them choose from a bar graph or a line graph (a picture graph is also possible if you can get pictures of the snack food).
6. Ask the student to explain what her graphed data says or what its importance is.

PIE CHARTS

Pie charts are often used to show relationships between parts of a whole. They can visually show different proportions. However, it is very difficult to see the difference between small fractions in a pie graph. For example, students may have a difficult time telling the difference between ⅙ and ⅐ on a pie chart. Students in general education math classes may need to be able to recognize what simple fractions look like on a pie graph. Parents or teachers can collect simple pie charts showing halves, thirds, and fourths and have the students identify what they represent. Using a circle to actually graph data is a more difficult task than will be taught in survival math.

GENERALIZATION ACTIVITIES

- Find examples of simple data charts and show them to the students. See if they can tell what the graph really says. The newspaper *USA Today* frequently has simple graphs that are easy to read.
- Have the student graph the number of minutes she talks on her cell phone each day to determine what the best plan for her is (i.e., how many minutes she needs a month).
- Graph information on sports teams (or individual players) you are following. For example, graph a player's home runs or points per game.
- Keep data on the student's performance in sporting or recreational activities such as bowling, basketball, or swimming. Make a graph to show how she is improving (scores in bowling, swimming times, etc.)
- Buy 3 or more different brands of microwave popcorn and graph how many unpopped kernels there are in each brand.
- Have the student divide a real pie or pizza into half and then cut each half into halves or thirds and discuss the fractions that are made.

CHAPTER 17

Decimal Fractions and Percent

Questions to be answered:

Can the student:

1. Visualize simple decimal fractions by using a hundredths chart.

2. Demonstrate that two places to the right of the decimal point are needed when you are working with money.

3. Demonstrate understanding of decimal mixed numbers by coloring appropriately the decimal mixed numbers on the hundredths charts.

4. Change a two-place decimal to a percent and a percent to a two-place decimal.

5. State the fractional equivalents for .25, 25%, .50, 50%, .75, 75%

6. Explain "percent off" in sales.

Chapter 11 covered fractions, essentially as they are needed in baking. A fraction is one form of a number that shows part of a whole, and is encountered most frequently in the context of cooking. Since the focus of this book is on practical uses of math, commonly used fractions such as ½, ⅓, ¼, ¾, and ⅔ were learned in the context of cooking.

Decimal fractions and percents are the other forms of numbers that show parts of a whole. Percents show parts out of a hundred, and decimals show parts of tenths or units of hundredths, thousandths, etc. In this chapter, we will again focus on uses of decimals and percents in the real world (primarily in shopping), as well as on understanding the relationships amongst common fractions, decimals, and percentages, which is a valuable survival skill (and also expected of students who are included in the general math curriculum).

Decimal Fractions

Fractions based on the number 10 or ordered in units of 10 are called decimal fractions. Decimal fractions can be written in our number system as tenths, hundredths, thousandths, etc. without showing the fraction form of one number on top of the other. For example, ½ can be written as .5, .50, .500, etc. in the decimal system. (It is assumed that the denominator is some unit of 10.)

Often it is easier to add, subtract, multiply, and divide using decimal fractions than using regular fractions (e.g., ½ x 20 vs. .5 x 20). This is especially the case when using a calculator. In addition, since our money system is based on decimal fractions, learning about them at least to the hundredths becomes a survival skill (so they can understand money amounts such as $23.88).

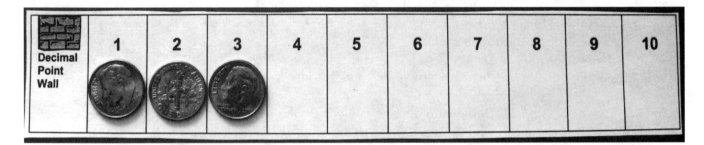

INTRODUCTION TO DECIMAL FRACTIONS—*TENTHS*

OBJECTIVE: The student will be able to visualize and reproduce simple decimal fractions by using a tenths strip and concrete tokens.

MATERIALS:
- 3 Decimal Tenths strips for each student
- Crayons or colored pencils
- 10 dimes
- Several colors of poker chips or counting chips for each student

SUCCESS STEP: Ask each student to count their pennies.

PROCEDURE:
1. Put a dime on one of the squares on the Tenths strip. Ask the students what fraction you have filled in. Help them to see that you have filled in 1 square out of 10, or one tenth. Have the students repeat the action with their own dimes and Tenths strip.
2. Ask the students how to write ¹⁄₁₀. Now explain that there is another way to write ¹⁄₁₀, namely **. 1**.

3. Put 2 dimes or tokens on 2 squares. Explain that the decimal fraction .2 is equal to $\frac{2}{10}$ or 2 parts of 10. It is read as *2 tenths* (or sometimes as *Point 2)*. Have the students put 2 dimes on their own strip

4. Put 3 dimes on the Decimal Tenths strip and see if the students can name the fraction and write it two ways.

5. Fill in more squares and ask the students to repeat the actions and to name the corresponding common and decimal fractions. Be sure that the students notice that the decimal numbers are to the right of the decimal point wall rather than to the left (which is for whole numbers).

6. Have the students color in the squares for $\frac{3}{10}$, $\frac{4}{10}$, and $\frac{1}{10}$—each with a different color.

7. Repeat the above steps (except for the coloring) with the colored counting chips.

INTRODUCTION TO DECIMAL FRACTIONS—*HUNDREDTHS*

It is important to teach the students about hundredths because that concept relates so much to both money and percent. For Survival Math, we do not have to go any farther than tenths and hundredths.

OBJECTIVE: The student will be able to show hundredths by correctly using a hundredths chart and writing the appropriate fraction and decimal fraction.

MATERIALS:
- Hundredths chart (Appendix B)
- 100 pennies for the teacher (optional)
- 10 Decimal Tenths strips for teacher
- 30 pennies for each student
- Chalkboard and chalk or whiteboard and marker

SUCCESS STEP: Show the tenths strip from the previous lesson and ask the student to point out $\frac{1}{10}$ (.10) and $\frac{3}{10}$ (.3)

PROCEDURE:
1. Tell students there are 100 squares on the hundredths chart. Put a penny on one square (or color one square in) and ask what fraction you have colored in. Get them to say that you have covered $\frac{1}{100}$, or one part out of 100.

2. Ask them how to write $\frac{1}{100}$ as a fraction. Then tell them there is another way to write $\frac{1}{100}$: as the decimal fraction .01.

3. Show the decimal hundredths chart.

4. Tell the students that the decimal fraction .21 is equal to 21 parts out of a hundred ($\frac{21}{100}$) and is read 21 hundredths. Show on the chart by counting out 21 squares.

5. Let the students put pennies on the various numbers (up to 30) of hundredths on the chart and name them. (The students (and teacher) may have to put all their pennies together if they want to show decimal fractions above 30.)

6. Explicitly teach the students how to write the decimals for .01 to .09. Put 1 to 9 pennies on the first row and ask them what fraction you have made and how to write it

two ways. If they write the decimal without the 0 in the tenth place (e.g., .3 instead of .03) ask them then how would they write 3 tenths? If they can't remember, get out the tenths strip from the previous activity and refresh their memories. Tell them that when there is only 1 number right after the decimal point, it is 10ths; you need 2 numbers after the decimal point for hundredths.

One Whole

	1	2	3	4	5	6	7	8	9	10
										20
										30
										40
										50
										60
										70
										80
										90
										100

RELATING DECIMAL FRACTIONS TO MONEY

OBJECTIVE: The student will be able to demonstrate how decimal fractions relate to the notation of money amounts.

MATERIALS:
- 100 pennies
- Hundredths chart (Appendix B)
- Comparing decimals (Appendix B)
- Whiteboard and marker or paper and pencil

PROCEDURE:
1. Remind the students that there are 100 pennies in one dollar.
2. Ask them if they know what fraction 1 penny is out of 100.
3. If they can't say, put 100 pennies on the hundredths chart. Emphasize that the 100 pennies on the chart equal one dollar.
4. Pick up one of the 100 pennies and ask, "How many pennies out of one dollar is this?" (One.) "It is one out of how many pennies?" (100.) So, one penny is $\frac{1}{100}$ of a dollar.
5. Ask them how you would write one penny in decimal notation. Remind them that it is $\frac{1}{100}$. Point out that this is how they have been writing cents in money notation. Cents are how many pennies (or 100ths of a dollar) you have.
6. Take one penny off the chart, leaving 99. Ask the students how many pennies are on the chart. (Don't let them count if they don't know automatically. Point out that you started out with 100, and took away 1…. What do they get if they count back 1 from 100?)
7. After they say 99, ask what fraction that is, and ask them how to write that as a decimal (.99).
8. Ask how many cents that is.
9. Put the missing penny back, so you have 100 pennies on the chart. Ask what fraction/decimal (or how much money) they have now. If they say "100 cents" ask if there's another way to say that. Guide them to say 1 dollar.
10. Ask how to write 1 dollar. Help them see that $\frac{100}{100}$ is 1.00, or $1.00, when dealing with money.
11. Ask the students how you would write the 30 pennies that they have in decimal form. Remind them that 100 pennies make one dollar. See if they can tell you that the fraction is $\frac{30}{100}$ or .30.
12. Using their pennies, see if they can make various decimals below 30 and read them correctly.
13. Remind them that when working with money, people always want to know how many pennies there are so you have to have ***two places to the right of the decimal point when you are working with money.*** So, .5 is the same as .50.
14. Use the Comparing Decimals worksheet from Appendix B. Cut out the Tenths strips, then lay them out so they cover the rows on the hundredths chart above. Show the students that you have now divided the chart into 10ths. Each strip is one-tenth.
15. Ask them how to represent $\frac{1}{10}$, $\frac{5}{10}$, etc. as a decimal. They should say .1. Then lift one of the Tenths strips and let them see that there are $\frac{10}{100}$s underneath. ***When we are working with money, we always want to have two decimal places to represent cents (unless there are no cents—.00).*** Then ask them to guess what 5/10s and 6/10s are—in tenths and then in hundredths.
16. Remind them that most calculators eliminate an ending zero, so something like .20 becomes .2. However, in money we need two places, even if the end number is a zero (.4 = .40 cents). If that issue becomes a stumbling block for a student, you can use PCI's Money Calc (www.pcieducation.com) or buy a calculator that can be set to only give 2 places in the answer (Texas Instruments TI -15 and others).

Comparison of Decimals

Many students have difficulty understanding the difference between 100 and .01. Hearing the difference between the words *tens* and *tenths* and *hundreds* and *hundredths* is difficult for anyone. When I am first teaching about tenths and hundredths, I like to pronounce tenths as "tents" and hundreds as "hundred(t)ents" to help the students hear the difference. I also like to draw little pictures of tents in the tenths and hundredths columns of place value charts to help them visualize the difference—see Place Value (Fat Mat) in Appendix B. What is most important is that the student understand the value of the decimal fractions.

INTRODUCING THE CONCEPT

OBJECTIVE: The student will be able to demonstrate the difference in value between 1 whole, 1 tenth, and 1 hundredth.

MATERIALS:
- Squares A, B, & C cut out from Comparing Decimals worksheet (Appendix B) for each student
- Place Value Fat Mat from Appendix B

SUCCESS STEP: Ask the student to name some decimals (fractions). See if his answers demonstrate he understands what decimals are.

PROCEDURE:
1. Ask the students to name what the shaded parts of A, B, & C are.
2. Pretend that the shaded parts are made out of real gold. Ask which one of the squares they would like to have.
3. Have the students answer the questions from the Comparing Decimals worksheet. Have them put the squares in order by value. Note that questions the teacher should ask are deliberately wrong. The students should know that 1 tenth is the same as 10 hundreds.
4. Give the students a Place Value Fat Mat from Appendix B. Review what the names on the Mat mean. Read some of the following money amounts out loud and see if the students can write the numbers in the correct places on the Place Value Fat Mat chart.

$.33 (Read the number as 33 cents)	$.49
$.01 (read as 1 cent)	$.09
$.99	$.15

Decimal Mixed Numbers

Students will also need to use decimal amounts that include whole numbers. These numbers are called *mixed numbers*. Indeed, they have already been using mixed numbers when they talk about dollars (whole numbers) and cents (decimal fractions).

COLORING MIXED NUMBERS

OBJECTIVE: The student will demonstrate understanding of decimal mixed numbers by coloring appropriately the decimal mixed numbers on the hundredths charts.

MATERIALS:

- Hundredths charts (two for each student)
- Crayons or markers
- Decimal Mixed Numbers Hundredths chart given below
- Place Value Fat Mat–Money

PROCEDURE:

1. Tell the students that the Mixed Decimal chart represents 1 (the whole number) and 8 hundredths, and is written 1.08. Note that the 8 is in the hundredths place with a 0 in the tenths place. If the number was written 1.8, it would mean 1 and 80 hundredths and the chart would be shaded down 8 rows.
2. Have the students color in 1.45 of the 2 hundredth charts that were given to them.
3. Have students practice naming mixed numbers out loud, saying "and" when they get to the decimal place. For example:

$$3.28 = 3 \ \underline{\textbf{and}} \ \text{twenty-eight hundredths}$$
$$2.05 = 2 \ \underline{\textbf{and}} \ 5 \ \text{hundredths}$$
$$7.9 = \ 7 \ \underline{\textbf{and}} \ 9 \ \text{tenths.}$$

You can supply as many mixed numbers as needed to understand the concept.

4. See if they can write down or punch into the calculator mixed numbers when you read them aloud—emphasize that they should press the decimal point when you say "and."

Decimal Mixed Numbers Hundredths Chart

5. Using the Place Value Fat Mat for money (Appendix B), have the students write down some of the following money amounts which include mixed numbers. The numbers are:

$1.44	$ .71	$2.50
$ .04	$ .08	$3.30
$ .18	$5.35	$24.65

Note: Be sure the students know that when they write down prices to add or subtract, the decimal point of one needs to match up with the decimal point of another:

$1.81
$ .22

Percent

Percent is a part of a whole that is divided into 100 parts, just like the hundredths discussed above. Percent actually means per hundred. I tell students that the sign for percent is %, which takes the 1 in 100 and puts it between the two zeros that make up 100. So, 1 0 0 becomes 0 1 0, with the 1 written as a slanted line: %.

INTRODUCING THE PERCENT SIGN

OBJECTIVE: The student will be able to recognize the percent sign and read percentages when notated with numerals and the percent sign.

MATERIALS:
- Whiteboard or paper with the percent symbol written on it
- Several newspaper ads showing sales as a percent off
- Packaged food with nutrition labels (percent of daily requirements)
- School papers with percent correct (one paper having 100%), if appropriate.

SUCCESS STEP: Show the students the percent symbol. Ask if they can see any numbers in the symbol. Praise answers of a 1 and two 0's.

PROCEDURE:
1. Holding up the sign, ask them what this sign is (percent). Ask where they have seen one before (on their tests? in stores? on boxes of food?).
2. Show a school paper with the score of 100%. Ask them if they got a 100% on a test, would that be a good or a bad thing? Why? What does it mean? See if they know that it means you got everything—the whole thing—right. 100% is another way of saying the whole thing, or one whole thing.
3. Ask the students if you had less than 100%, would that be more or less than 1 whole thing? (less)
4. Tell the students that the word percent actually means per hundred. Tell them that the sign for percent is %, which takes the 1 in 100 and puts it between the two zeros that make up 100. 1 0 0 becomes 0 1 0, only the 1 is written as a slanted line: %.

5. Have the students write some percents that you call out loud (on paper or slates).

31%	28%	33%	4%
3%	11%	9%	5%

6. Pass around some newspaper ads or nutrition labels and ask the students to find the percent signs and read the percentages aloud.

CHANGING PERCENTS TO DECIMALS

OBJECTIVE: The student will be able to change a two-place decimal to a percent and a percent to a two-place decimal.

MATERIALS:
- 100s Charts from previous discussion above
- Pencil and paper
- Percents to Decimals Worksheet #1

SUCCESS STEP: Ask the student to put a big percent sign at the top of his paper. Help him to do it correctly, if necessary.

PROCEDURE:
1. Use the previous chart to remind the students that the chart shows **.21** and is read **twenty-one hundredths.** Remind them that **.21 hundredths means 21 parts of a whole number 1.** Since percent really means hundredths, .21 is the same as **21%.**
2. Explain that any number percent can also be written as a decimal with two places to the right of the decimal point. For example, 68% = .68 as a decimal. And .68 as a decimal can be changed into 68%.
3. Tell the students that the tricky part comes when the percent has only one digit—like 3%. One-digit percents must still be changed to two-digit decimals. So, 3% becomes the decimal .03, not .30. The same is true in reverse. One-digit decimals must be changed to two-digit percents, so .3 becomes 30%.
4. Go through the following problems with the students orally, explaining each problem and having the students name the resulting percents and decimal fractions.

Change each Percent to a Decimal:

33% _____   23% _____   8% _____   18% _____

3% _____   27% _____   12% _____   5% _____

5. Have the students use the 100s chart for assistance. For instance, if they change 3% to .3, show them that you would cover 3 squares for the one, but 30 squares for the other.

Remember—Percent means hundredths. If you have **2%**, you write it as the decimal **.02**, i.e., when percents are changed into decimals, they always have **2 places to the left of the decimal point (except for 100%, which is 1).**

6. Have the students complete the Percents to Decimals Worksheet #1 independently.

MATCH 'EM GAME

OBJECTIVE: The student will be able to match common percents with corresponding decimal fractions in a game setting.

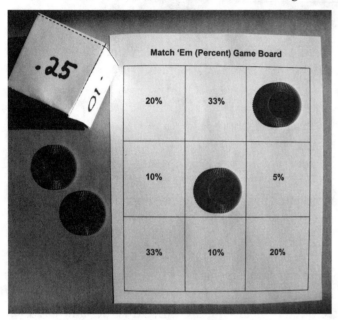

Match 'Em (Percent) Game Board

MATERIALS:
- "Custom-Made" Dice from pattern in Appendix B24. Mark the die with the following numbers: .01, .25, .50, .20, .10, .33
- Match'em game board from Appendix B101
- 10 tokens or bingo chips for each player.
- Paper and pencil or slate board and chalk for each student

PROCEDURE:
1. Have the students read the decimals on the dice out loud. Put the game board where all players can see.
2. The first player rolls the die. He covers the square on the game board that corresponds to the die that he rolled.
3. The play goes to the next player. If the die comes to a square that is already covered, the next player takes his turn.
4. The player who covers all nine squares first is the winner.

USE OF PERCENT

Your student(s) will quickly see the usefulness of being able to convert percent into decimals when he buys something on sale at a certain percent off the regular price. To find out how much you save, you convert the percent off into a decimal fraction and multiply the decimal times the original price. The answer is the amount of money that you can subtract from the original price to get the price you pay.

OBJECTIVE: The student will learn to convert percents to decimals for the purpose of understanding "percent off" sales.

MATERIALS:
- Paper and pencil or whiteboard and marker
- Percents to Decimals Worksheet #2

PROCEDURE:

1. Show the students how to work through the following example on paper or a white-board or slate.

> # 60% <u>OFF</u>
> ## SALE
> **Mountain Bike**
> Original price $200

What is the sale price of the bike?

- Convert 60% to .60
- Multiply .60 x the original price of the bike ($120).
  This tells you how much money you save by buying the item on sale.
- Subtract the amount of the discount from the original price
  ($200 – $120 = $80.)
- $80 is the amount you will pay for the mountain bike.

2. Help the students do several problems of this type. Then have them do the Percents to Decimals Worksheet #2.

Using a Calculator to Calculate Percent Off

When the students have some experience with converting decimal fractions to percents, tell them that if they are using a calculator when shopping, they can use a shortcut way to find the final price. Have the students do the Percent to Decimal Worksheet using the calculator shortcut.

- Enter the original price of the item into the calculator ($200)
- Enter the subtraction sign (-)
- Enter the amount of the sale percent (60)
- Enter % sign
- Answer is the amount of money you will have to pay ($80).

It is important for the students to first understand how to figure out the final price by converting the percent to a decimal and subtracting from the price because the shortcut calculator way doesn't explain what really happens. Once the students understand percent off, they can use this quicker way, if they are carrying a calculator.

ALTERNATE WAYS OF HANDLING PERCENT WHEN SHOPPING

Many students may not be able to figure out the sale price of an item when shopping. In fact, many typical adults can't do the mental math that it takes to figure out the sale price. Realizing this, many stores just put the sale price on the signs or tickets on the item.

If students carry their calculators with them when shopping, the most accurate way is to use the calculator shortcut explained above. However, some young adults I know have worked out an alternative way, which, though not as accurate as figuring

out the exact amount of the discount, can give them an estimate of the sale price (if they don't carry their calculators.) These young adults figure that:
- 50% off really means that the item is half price.
- 40% off is a little less than half (50%) off,
- 60% off is a little more than half off (50%).
- They have learned to divide numbers by 2 in their heads. Therefore, they just divide the original price in half and figure that they will have to pay a little more if it is 40% off, or a little bit less if it is 60% off. One of them said, "It isn't really a sale if there is less than 40% off."

This method seems to work for them and who am I to argue with success? I think that they just use it when they are estimating the price to see if they can afford something. When they really have to cough up the money, they use their calculators. It seems that it would pay to work with your students on dividing by 2 in their heads, so they could use those facts in survival math.

Hints on Percents

Tips

I don't teach a method for figuring tips, especially for sit-down restaurants. Instead, I recommend teaching students to use a tip card. Tips cards for several percentages are easily available from card stores, drug stores, etc. Students can carry a card in a wallet or purse—wherever they carry their money.

Ads

Most newspaper ads list sales as a percent *off*—e.g., 15% **off** of $100, which equals $85. Sometimes the ads say 15% **of** the original price, which in the case of $100, is $15. 15% of (times x) the original price of $100 means a much smaller price ($15) than 15% off. You just have to emphasize that **off** means an amount off or *subtracted* from the price. **Of** means times, which is *multiplication*.

SOME COMMON DECIMAL EQUIVALENTS OF FRACTIONS

For some hands-on learners, it is also handy to memorize the most common percentage and decimal equivalents of fractions. Knowing these equivalents can be especially useful when shopping.

OBJECTIVE: The student will be able to mentally convert a few common fractions to decimal fractions and to percents.

MATERIALS:
- Flap book that has fractions on the top flaps and decimals and percents on the corresponding bottom sections (instructions and model are in Appendix B)
- Four quarters; dollar bill

SUCCESS STEP: Show the student a quarter and ask him how many quarters make 1 dollar. Praise a correct answer or show him four quarters alongside of a dollar to get the right answer. Point out that a quarter is 1/4 of a dollar. Tell him that that we are going to learn what the decimal fractions or percents are for common fractions like ¼.

PROCEDURE:
1. Show the students the flap chart and explain how to use it. Have them make a similar chart for their own use, if possible.
2. Have the students practice looking at the top number and trying to say the bottom decimal and percent. They can keep testing themselves as they learn.

GO FISH GAME

OBJECTIVE: The student will be able to match a common fraction with its equivalent decimal fraction and percentage in a game context.

MATERIALS:
- Go Fish cards from Appendix B copied to cardstock (or similar cards made from index cards)
- Flap book from previous activity

PROCEDURE:
1. Shuffle cards. Deal four cards to each player (two or three players). Put the remaining cards face down in between the players.
2. The first player looks at his cards and asks one of the other players for a card that will match one of the cards in his hand. For instance, if he wants a match for ½ he asks, "Do you have anything that is worth one-half?" When they are first learning to play, players may look at the Flap chart above to help in learning equivalents.
3. If that person does not have an equivalent card (e.g., 50% or .50), he tells the first player to *Go Fish:* pick a card from the pile of cards in the center.
4. The next player then asks for a match for one of his cards.
5. When the player has all 3 of the same value cards, he puts them face down together by his place.
6. The winner is the person with the most matches when all the cards in the center are gone, players can no longer obtain matching cards from other players, or when a specified amount of time has elapsed.

GENERALIZATION ACTIVITIES
- Find library books on the shelf using the Dewey Decimal system (e.g., 526.77).
- When grocery shopping, show your child how you read the labels for products sold by weight. For example, show him that you are buying a pound and a half of ground beef or coffee beans, but that the label says 1.50 lbs. Or have him help you look for a quarter-pound package of cheese.
- Use a pedometer or a bike odometer that has a digital readout. Talk about whether you have walked just a little more than 1 mile, or close to a mile and a half, or nearly 2 miles. (Make a graph of how far you walk each day.)

- Read the temperature on a digital fever thermometer.
- Discuss baseball or other sports statistics that are expressed in decimal form in the newspaper.
- Help your child understand in a general way the percent bars you see when using the computer, especially on the Internet (e.g., you have to wait until something is 100% loaded to play, and the line keeps progressing across the screen, getting gradually longer, as the percentage loaded increases). Or when you are downloading a picture, talk about the percent bar that shows how much of the picture you've downloaded so far.
- If your child uses a digital camera, show him how to size it up or down on the computer screen (or when using a photo developing kiosks in a store) by percentage. If you make it 50%, the picture will be half the size.
- If your child/student uses a photocopier, show him how he can use percentages to make the print size bigger or smaller.
- Help your child understand in a general way what percent scores on his tests mean.
- Teach an older child or adult about tipping 15% or 20% (using a tip card).
- When you see an interesting pie chart, such as in *USA Today,* help the student understand what it means, emphasizing the percentages he knows fractional equivalents of. For example, if the pie chart shows that 55% of people prefer chocolate ice cream, is that more than half (50%)?
- If your older child is interested, look at nutrition labels together when you are shopping. Talk about whether it is a good idea to buy the ice cream that has 65% RDA of fat per serving or the one with 25%? Should you buy the brand of soup that has 80% RDA of sodium? or 30%?
- Discuss the weather forecast. If there is an 80% chance of rain, would you be wise to bring an umbrella? If there is 90% humidity, would it be a good day to go running?

CHAPTER 18

Story Problems

Questions to be answered:

Can the student:

1. Reword story problem questions as answer statements with a blank.

2. Draw or act out the problem situation.

3. Determine the operation that should be used.

4. Write a number sentence for the problem.

5. Calculate the answer (with or without a calculator).

6. Write the answer in the blank of the answer sentence.

7. Check the problem.

8. Figure the average (mean score) of a group of scores.

To apply the math skills they learn in the classroom to their own lives, students need to have real-life and simulated experiences with solving problems using math. This book will cover situations simulating actual problems, but the important learning must occur when students use math in their own lives. Parents and teachers need to furnish many opportunities for students to use their math skills for solving real problems. The structured approach for word problems described here has been explained in Chapters 5, 9, and 12. This chapter summarizes how to use the structured approach when all four operations (addition, subtraction, multiplication, and division) have already been mastered.

It is most important that the student understand what the problem is asking. In classroom settings, students often get confused by the language that is used. The first step needs to be to read the problem and then paraphrase it into simpler language. At first the teacher or facilitator will probably need to explain the problem simply. She may possibly need to demonstrate what is needed by acting out the problem with manipulatives or by making drawings.

For example:
Mrs. Hall's class had an election for class president. Ted received 11 votes. Marsha got 13 votes, and Michael received 7 votes. How many students voted in the election?

- **Read the problem.**

- **Reword it in your own words.**
  Voting for class president: Ted-11 votes, Marsha-13, Michael-7.
  (Each person could only vote once.)

- **Reword each question as a statement.**
  The total number of votes was ____ .
  (When you reword the problem, write what the final sentence will be,
  leaving the number blank)

- **Draw or act out the problem, if possible.**

Ted	Marsha	Michael

Total Votes = ____

- **Decide what operation will be used.**
  ○ Is the answer going to be bigger than any of the numbers given in the problem?
  Yes, the total amount of votes will be bigger than the number of votes each person received votes. Therefore, the operation must be addition or multiplication. If it is multiplication, the number of items in the groups you are combining need to be the same (equal). The numbers (11, 13, 7) are not the same, so the operation must be addition.

- **Write the number sentence.**

$$11 + 13 + 7 = \underline{\quad}$$

- **Solve the problem, using a calculator, if necessary.**
  The total number of votes was 31.

- **Check your answer. You can check the answer by counting the marks that were drawn** or by adding the numbers in the reverse order.

More Detail on Problem-Solving Steps

REWORDING THE PROBLEM

In rewording the problem, the student should put the problem in her own words, writing down the important numbers. One essential step in rewording should be to write what the final sentence will be in regular word order (for adults: subject, verb, object) not in a question or using question words such as *how many* or *what*. Essentially the question ending with a question mark (?) is changed into a statement ending with a period (.) At this time the student leaves a blank space where the answer will be.

Story problems often have a question beginning with *how many, how much, how far, what is,* where the subject (person or object) is not at the front of the sentence. Many students with Down syndrome or other disabilities that affect communication skills have difficulty finding out what the problem is really asking for because of the transformation in grammar that occurs when we ask a question. At first, the teacher will have to supply the final sentence, but eventually the student should be able to reword the problem into a simple subject-verb type complete sentence.

For example:

- Paul studied his spelling words for 24 minutes. He studied his math flashcards for 13 minutes. How long did he study?

  The final sentence could be reworded as:
  Paul studied for _____ minutes.

- Jake bought 2 packages of M&M's. They each cost $.59. How much did he pay for both of the packages of M&M's?

  The final sentence could be reworded as:
  Jake paid _____ for two packages of M&M's.

- Emily bought two hot dogs for $1.50 each. How much did she pay for the hot dogs?

  The final sentence could be reworded as:
  Emily paid $_____ for the hot dogs.

I usually have the student write the numbers from the problem as part of the rewording because then the important parts are all together in the story problem frame. Sometimes story problems have extra numbers that are not needed to solve the problem. (Luckily, this does not happen much in real life. When we are solving a real problem, we usually choose only the numbers that we need to solve the problem.) Tell the students that test makers may sometimes try to trick people by adding extra numbers that are not needed. When the students reword the problem in their own words, they should ask themselves if all the numbers are going to be used to solve the problem.

For example:

- Emily used her $5 bill to buy two hot dogs for $1.50 each. How much did she pay for the hot dogs?

   The problem can be reworded as:
   Emily paid $_____ for the hot dogs.

   Do we need to know what kind of bill Emily used to pay for the hot dogs?
   No—-they just threw in an extra number to confuse us.

CHOOSING
THE OPERATION
TO BE USED

Many students have difficulty deciding what operation to use, especially when they have learned all four operations of addition, subtraction, multiplication, and division. They often want someone else to just tell them what operation to use. Some students add the first two numbers, whatever they are, and others tell me that they just use the operation that they have been using in class before being assigned the problems.

It is important that students actually think through the individual problem (single-step problem) and decide themselves what operation to use. It is important that the students think through the story problem, including the possible answer. There are two factors that need to be determined for *whole number* problems:

1. *Is the answer going to be larger than any of the numbers given in the problem?* If a total of various numbers is to be found, then the answer will be larger than any of the single numbers involved. If the answer is going to be larger than the numbers given, the operation must be **addition** or **multiplication**. If the answer is going to be **smaller** than at least one of the numbers given, the operation must be **subtraction or division.**

   For example:

   - Declan has $12. Bonnie has $7. How many more dollars does Declan have?

      Reworded as:
      Declan has _____ more dollars than Bonnie.

   Declan can't have more dollars than either of the two amounts of $12 or $7, so the operation is either subtraction or division.

2. *Are the numbers given the same or equal to each other?* Statements such as **each** boy had 6 marbles or that 20 cookies need to be divided **equally** between 5 children indicate the numbers are equal. **If the numbers are the same, you are combining equal groups or dividing something into equal groups. Then the operation must be multiplication or division.** Knowing that the numbers are equal or not and whether the answer should be larger or smaller than the numbers given gives the proper operation.

NOT Equal Numbers		Equal Numbers
Answer Larger ⟶	ADDITION	MULTIPLICATION
Answer Smaller ⟶	SUBTRACTION	DIVISION

Available as a sign/cue sheet in Appendix B

PRACTICE CHOOSING THE OPERATION

OBJECTIVE: The student will be able to choose the operation in simple, one-step story problems.

MATERIALS:
- Oat cereal such as Cheerios
- Construction paper (1 piece per student)

SUCCESS STEP: Ask your student(s) to count out 20 pieces of oat cereal and put them on a piece of construction paper.

PROCEDURE:
1. Act out the following problems:
   - Call out a student's name and remind her that she has 20 pieces of oat cereal. Ask her to separate them into 2 *equal* piles.
   - Don't have her count how many pieces are in each pile. Instead, ask her the following questions to walk her through how to determine the operation (even though it would be easy to figure out the answer without thinking about the operation).
     - **Is the answer going to be larger than the biggest number in the problem?** No, because each pile wouldn't be larger than the total number of pieces of oat cereal. Therefore, the operation must be subtraction or division.
     - **Is the number of oat pieces going to be equal in each pile?** Yes, the problem itself says that the piles are equal numbers, so the operation would have to be division. You have then figured out what the operation should be.

2. Write the numbers on the form on the next page:
   Problem: Divide 20 Cheerios into 2 equal piles. How many in each pile?

READ AND UNDERSTAND THE PROBLEM	DRAW, USE NUMBER LINE, OR ACT OUT PROBLEM
(Write needed numbers with their labels; e.g., 4 desks or underline them in the problem) 20 Cheerios 2 piles *(Write answer sentence with a blank)* The total in each group is _____ Cheerios. 1	O O O O O O O O O O O O O O O O O O O O 2

FIND OPERATION:	CIRCLE OPERATION
Will the answer be larger than the biggest number in the problem? (No) Are the numbers in the problem the same or equal? (Yes) 3	<table><tr><td></td><td>equal groups ⬇</td></tr><tr><td>Larger ➜ Addition</td><td>Multiplication</td></tr><tr><td>Smaller ➜ Subtraction</td><td>Division</td></tr></table> 4

WRITE NUMBER SENTENCE (e.g., 8 + 4 = ___)	WRITE ANSWER SENTENCE
20 divided by 2 = _____ **SOLVE PROBLEM (e.g., 8 + 4 = 12)** 20 divided by 2 = 10 5	The total in each group is 10 Cheerios. **CHECK THE PROBLEM** $10 \times 2 = 20$ 6

3. Next, have the student divide the Cheerios among all the people present, thus making a new similar problem and asking the same questions. (If the numbers do not come out evenly, you take the remainder so the piles are equal.)

4. Fill in the blank form below with the following story problem:

Alice sends postcards to her friends while she is on vacation. She bought 16 postcards. She wants to send them to 4 friends. If she sends each friend the same number of postcards, how many postcards will each friend receive?

READ AND UNDERSTAND THE PROBLEM

(Write needed numbers with their labels; e.g., 4 desks or underline them in the problem)

16 postcards   4 friends

(Write answer sentence with a blank)

Each friend received _____ postcards.

1

DRAW, USE NUMBER LINE, OR ACT OUT PROBLEM

| 1 | 2 | 3 | 4 |

2

FIND OPERATION:

Will the answer be larger than the biggest number in the problem? (No)
(Each friend can't get more than the total amount of postcards Alice bought.)

Are the numbers in the problem the same or equal? (Yes)

3

CIRCLE OPERATION

	equal groups ⬇
Larger ➜ Addition	Multiplication
Smaller ➜ Subtraction	Division

4

WRITE NUMBER SENTENCE (e.g., 8 + 4 = ___)

$16 \div 4 =$ _____

SOLVE PROBLEM (e.g., 8 + 4 = 12)

$16 \div 4 = 4$

5

WRITE ANSWER SENTENCE

Each friend received 4 postcards.

CHECK THE PROBLEM

$4 \times 4 = 16$

6

5. Have the students work out the first problem, step-by-step, with you. At first, it might be best to act out the problems. Depending on the number of people you have available, you may want to act it out using Post-It notes, as described below. Then let them do two more problems and correct those problems orally. Let them do the additional problems on the following pages. All the problems will come out even.

DETERMINING THE PROCESS NEEDED
Essentially, the process for deciding the operation is:
- If the answer is going to be **larger (or more)** than the largest number in the problem, we use **addition or multiplication.** Conversely if the answer is going to be **smaller (less), than the largest number** in the problem, we use **subtraction or division.**
- **How do we decide between addition or multiplication?** If the problem indicates that **equal** amounts are to be given, the process is multiplication. If not, the process is addition. (The words **same, each,** and **equal** are often part of the word problem when multiplication is needed.)
- **How do we decide between subtraction or division?** If the total amount is going to be separated into **equal groups,** the process is division. If not, the process is **subtraction.**

ACTING OUT STORY PROBLEMS

OBJECTIVE: The student will be able to demonstrate understanding of simple, one-step story problems by acting out the problem situation.

MATERIALS:
- Manipulatives (depending on the items in the problems)
- Two different colors of small Post-It notes

SUCCESS STEP: Have the student put out on the table the number of Post-It notes that are appropriate to the beginning of the problem.

PROCEDURE:
1. The teacher writes out a label on each of the Post-It notes that the student has put on the table. For example:

- Emily and her family went out to eat at a restaurant. All 6 members of her family ate the same thing (equal amounts of food). The total bill for the meal was $30. The cost of each person's meal was $_____.

In the above problem, 6 Post-It notes would be labeled Emily, Father, Mother, Brother, Brother, and Sister. The total bill is $30, which will be shown with 30 single dollar bills or 30 small Post-It notes in a different color than those representing the family members.

The student "deals out" the single dollars or Post-It notes so that each family member has 6 notes or dollars.

READ AND UNDERSTAND THE PROBLEM

(Write needed numbers with their labels; e.g., 4 desks or underline them in the problem)

6 members in family
Total bill is $30.

(Write answer sentence with a blank)

Each family member spent _____ on dinner.

1

DRAW, USE NUMBER LINE, OR ACT OUT PROBLEM

1 [money]
2 [money]
3 [money]
4 [money]
5 [money]
6 [money]

2

FIND OPERATION:

Will the answer be larger than the biggest number in the problem? (No)
(The price of each dinner can't be more than the total bill.)

Are the numbers in the problem the same or equal? (Yes, they each had the same thing—equal price)

3

CIRCLE OPERATION

	equal groups ↓
Larger ➜ Addition	Multiplication
Smaller ➜ Subtraction	Division

4

WRITE NUMBER SENTENCE (e.g., 8 + 4 = ___)

$30 \div 6 =$ _____

SOLVE PROBLEM (e.g., 8 + 4 = 12)

$30 \div 6 = 5$

5

WRITE ANSWER SENTENCE

Each family member spent $5 on dinner.

CHECK THE PROBLEM

$5 \times 6 = 30$

6

SAMPLE PROBLEM A

Moira found 15 spiders on the porch. Season found 7 spiders in the hall. How many spiders were in the house?

READ AND UNDERSTAND THE PROBLEM

(Write needed numbers with their labels; e.g., 4 desks or underline them in the problem)

15 spiders   7 spiders

(Write answer sentence with a blank)

The total number of spiders
in the house was _____.

1

DRAW, USE NUMBER LINE, OR ACT OUT PROBLEM

✳✳✳✳✳✳✳✳✳✳✳✳✳✳✳

✳✳✳✳✳✳✳

2

FIND OPERATION:

Will the answer be larger than the biggest number in the problem? (Yes)

Are the numbers in the problem the same or equal? (No)

3

CIRCLE OPERATION

	equal groups ⬇
Larger ➡ Addition	Multiplication
Smaller ➡ Subtraction	Division

4

WRITE NUMBER SENTENCE (e.g., 8 + 4 = ___)

5 + 7 = _____

SOLVE PROBLEM (e.g., 8 + 4 = 12)

_____ ☐ _____ = _____

5

WRITE ANSWER SENTENCE

(Write the answer in the blank of the answer sentence in square 1)

6

SAMPLE PROBLEM B
On their summer trip, the Barbers drove 55 miles per hour for 5 hours. How many miles did they drive?

READ AND UNDERSTAND THE PROBLEM	DRAW, USE NUMBER LINE, OR ACT OUT PROBLEM
(Write needed numbers with their labels; e.g., 4 desks or underline them in the problem) 55 miles per hour 5 hours *(Write answer sentence with a blank)* The Barbers drove _____ miles on their trip. 1	2

FIND OPERATION:	CIRCLE OPERATION
Will the answer be larger than the biggest number in the problem? (Yes) *(The total amount of miles driven will be larger than the number of miles per hour.)* Are the numbers in the problem the same or equal? (Yes) 3	(see operation table below) 4

CIRCLE OPERATION table:

	equal groups ↓
Larger ➜ Addition	Multiplication
Smaller ➜ Subtraction	Division

WRITE NUMBER SENTENCE (e.g., 8 + 4 = ___) SOLVE PROBLEM (e.g., 8 + 4 = 12)	WRITE ANSWER SENTENCE *(Write the answer in the blank of the answer sentence in square 1)*
_____ ☐ _____ = _____ 5	6

2. Sample problems that may be acted out:

- Paul planted 8 plants in the front yard. He then planted 6 more plants in the back yard. The total number of plants he planted is ____.

- Ryan picked up 14 shells on the beach. When he got home, 7 shells were broken. The total of unbroken shells was _____.

- Dad bought small ice cream cones for everyone in the family (8 in all). If each cone costs $1, the total cost for the family will be $____.

You can decide whether your student or students need more acting out of the problems. You might start acting out a few problems and then graduate to drawing the problem situation. When the students are not yet secure in doing story problems, they might be able to choose between acting out with Post-it notes and drawing the problem to make the situation concrete. The problems you ask them to solve should use small quantities so the acting or drawing will not take up too much time.

3. An option used by some problem solvers is to redo the problem using very small, simple numbers where it is easy to see what the answer will be. Then the same process is used to work with larger numbers. For example, a problem comparing $51.56 and $22.77 can be modeled by first comparing the difference between $5 and $2. Once the student understands what operation she needs to use with the smaller numbers, she can use that same operation with the larger numbers.

4. Have the students practice the sample story problems with your guidance as they go along. A few story problems are given here in the main text, but more problems are included in Appendix B.

More Practice with Word Problems

To give the students additional practice using the structured form:
1. Have the students do the Word Problem worksheet (Appendix B).
2. You may want to go through the problems orally with the students first.

They will need much more practice than can be given in this book. You can make up story problems for them. I have found that students are much more interested in solving word problems if I make up problems using their names and situations that are common in their lives. Eventually they will not need to use all of the steps in the structured form. However, it is important that they continue to look at the problems in the same way as done in the structured form with less written down. The essential steps are:

- Reword the question so it is a statement with a blank for the answer.
- Ask the two questions needed to figure out what operation to use.
- Write a number question and solve the problem.
- Put the answer in the blank of the answer sentence.

THE JOURNEY GAME

OBJECTIVE: The student will be able to do simple addition, subtraction, multiplication, and division number and word problems in a game context.

MATERIALS:
- The Journey game (as used in informal assessment), Appendix B
- Die
- Problem cards for addition, subtraction, multiplication, and division—both number and word problems (Appendix B)
- Game markers
- Calculator (optional)

PROCEDURE:
1. Place the game markers on Start.
2. One student throws the die. She moves her marker the number of dots she has thrown with the die.
3. The instructor hands her the appropriate card (addition, subtraction, multiplication, and division number and word problems). If the student can solve the problem, with or without the calculator, she moves the marker one place ahead if the problem is a number problem and two places if it is a word problem.
4. The players must follow the instructions that are given on the spaces where they land.
5. The winner must land on the finish by exactly the right number. She may also win by doing a problem correctly and moving ahead one or two places.

As the students get better at playing the game, you can decide whether you want to control what problems the student gets or you can put the appropriate cards face down on the game board and let the student choose. You can make new problems and use the Journey game over and over again.

FINDING THE AVERAGE (MEAN)

Mark has test scores of 87, 79, 92, and 86 at the end of the first quarter. The sum of the scores is 344. Dividing 344 by the number of test scores (4) gives the answer of 86. His teacher needs to have an average test score to figure his report card grade for the quarter. The score of 86 is a middle score that describes Mark's test scores for the quarter.

We are not going to discuss multi-step problems in general in this book. However, for many hands-on learners, it will be useful to understand the concept of average or mean. Even outside of math class, we often encounter averages in daily life—whether in relation to rainfall, sports statistics, prices, or grades.

The average is a measure that is considered the middle of a set of data. Frequently we need to add a group of numbers and then find out what the average is by dividing the total by the number of numbers in the set.

Finding the average is a multi-step process that requires you to:
1. Add the numbers together and get a total.
2. Count the number of numbers.
3. Divide the total by the number of numbers.

OBJECTIVE: The student will be able to find the average of four to six numbers.

MATERIALS:
- Three or four sets of materials such as pretzels, baseball cards, CD's, Legos, pencils, etc. (with unequal numbers of items in the sets)
- Calculators
- Paper and pencils
- Finding the Average worksheet from Appendix B

SUCCESS STEP: Ask the students to count the number of items in their sets.

PROCEDURE:
1. Ask the students or family members (you need to have at least 3 people) to put their items in front of them on the table. Count the number of items each person has and write the number of items on a paper near each item.
2. Ask the students to add the number of each set of items with their calculators.
3. Have the students count how many people there are.
4. Tell the students to divide the total number they got on their calculators by the number of people having sets.
5. Write the final number on a paper and label it the *average*.
6. Pass out the Finding the Average worksheet and work out the first two problems together with your student(s). Then have them try to solve the remaining problems by themselves. Watch them and help, as needed.
7. Look for opportunities in daily life to talk about *averages* that are meaningful to the student. For example, talk about the average number of points she scores per basketball game or the average amount of time people need to wait in line for a popular ride at an amusement park (assuming that information is posted somewhere).

GENERALIZATION ACTIVITIES

Students need to be able to solve story problems, but even more importantly, they need to solve real-life problems. Families and teachers need to be on the alert to pose problems requiring math solutions for the students to solve. In particular, look for real-life problems that the student has a stake in solving—that are meaningful to her. For instance, most students would be very motivated to learn how many slices of pizza they can have if everyone in the family gets an equal share. They would be less motivated to figure out how many ounces of detergent there are in the three different bottles on the shelf.

Most math books have only a few word problems compared to the number of pages of abstract math problems they have. You can find many more word problems by searching the Internet for "math word problems." Some websites that offer word problems at the time of this writing include: www.mathplaygound.com, www.mathcats.com, and www.mathstories.com.

However, doing more and more word problems is not the cure for difficulties with problem solving. It is more important to thoroughly discuss how to solve some problems than to do large quantities of problems. I also think that some story problems are designed to make the student fail. Many problems have extra numbers that are not needed. How often does that happen to you in real life? Not too often, I would guess. If you see a real situation in which there are too many numbers, you can deal with them at that time.

Probably the best way for students with Down syndrome and other disabilities to practice word problems is for you to make up real problems from everyday life for them. Parents can make up problems when they are shopping or putting away groceries. Teachers can make up problems when planning parties or trips. The way that students can really use math is by having real experiences that require math.

CHAPTER 19

Standards, Algebra, and Hands-on Math

Standards-based Math Instruction

In the United States, most states are now operating under standards for math that are very similar to those put out by the National Council of Teachers of Mathematics (NCTM) beginning in 1989 and more current revisions. The "No Child Left Behind" Elementary and Secondary Education Act of 2001 (NCLB) has mandated that states adopt similar standards and then conduct assessments based on these standards. These academic content standards provide a clear indication of what students should know at the various grade levels. Accountability for student learning according to these standards is accomplished by testing all students, except for a very few students with severe cognitive disabilities who use an alternate assessment. Most students' learning is organized around these math standards, even if they have disabilities and are still concrete learners.

These math standards are based on the following areas:
- Number and Operations Standard (number sense and computation)
- Measurement Standard (various attributes of items and how they are measured)
- Geometry Standard (two- and three-dimensional shapes and spatial concepts)
- Algebra Standard (understanding patterns; problem solving with algebraic symbols)
- Data Analysis and Probability Standard (collect, organize, and display data; probability)
- Mathematical Processes Standards (usually done while doing other standards)
    - Problem Solving
    - Reasoning and Proof

○ Communication
○ Connections
○ Representation

Many students with Down syndrome, autism, and other concrete thinkers can learn and use the basic principles of computation, with or without a calculator. Much of the survival skills in both *Book 1* and *Book 2* have focused on the standards of number sense and measurement, along with some geometry, problem solving, and data analysis.

Algebra (and Pre-algebra)

Parents and teachers have asked me about introducing algebra to students with Down syndrome and other hands-on learners. The truth is, if your child spends any amount of time in a typical classroom learning math, he will be exposed to pre-algebra and then algebra concepts without formerly being enrolled in an algebra or pre-algebra class.

In the younger grades, the algebra standard focuses on recognizing patterns and solving problems with an unknown quantity (e.g., $15 + n = 20$). In later middle school or high school, some students with Down syndrome, autism, etc. have been able to successfully participate in general education pre-algebra and algebra classes. Usually they are students who have understood the concepts of basic mathematics in earlier grades but have needed to use a calculator because of a memory problem with the basic math facts.

Students who have successfully mastered the lessons in this book may be ready to tackle pre-algebra or algebra. There are three different skills and practices that students using the *Teaching Math* books should have mastered that will assist them in understanding the "big ideas" in pre-algebra or algebra. The first is the frequent use of a number line to visually illustrate the basic operations. The concept of the number line can easily be extended to help hands-on learners develop an understanding of positive and negative numbers, which is needed for algebraic problem solving.

Second, both books discuss that one important step in the problem solving process is making a number sentence (which is really an equation) out of a word problem. Therefore, students are accustomed to seeing equations such as $2 + 6 = $ ____ and $2 + $ ____ $= 8$. It is only a short step to being able to use an "x" to represent the unknown quantity.

Third, some of the basic "big ideas" underlying basic operations and crucial to an understanding of algebra have been emphasized and illustrated in a hands-on manner. For example, the commutative property of addition (numbers can be added in *any* order) was taught and experienced from the very beginning of the instruction in addition (Book 1, p. 162). The fact that any number plus 0 (zero) equals itself and that any number times 0 equals 0 were also taught. The different uses for subtraction (take away and types of comparisons) were experienced and labeled. The concept that multiplication is repeated addition of equal numbers, and that division is repeated subtraction of equal numbers is illustrated and reinforced in this volume in the sections covering the structured form for solving word problems. Those underlying rules and principles also influence the understanding of algebra.

Guidelines for Participation in General Education Algebra

If you have a student who might be able to participate in a general algebra class, I suggest considering the following questions:

- Has the student been able to learn the useful math concepts that he will need in life after school? (If not, will he have opportunities later in his school career to work on consumer math skills?)
- Will the instruction in algebra class be adapted so he will be able to learn the "big ideas" of algebra and not be frustrated with acres of homework in things that will not have relevance in his life?
- Is the student himself motivated to put in the hard work needed to learn algebra, or is taking algebra more the parent's goal, not the student's?
- Is algebra necessary for getting a diploma for graduation? With the current federal laws, some states expect a student to keep up with the general education math requirements, including algebra, in order to graduate with a diploma. That includes passing the math achievement or graduation tests.

One mother told me that her daughter is in general education seventh grade math, and she's getting next to nothing out of it. They are doing algebraic equations with negative numbers, and the student hasn't a clue what they're doing and why. She is missing some of the basic prerequisites to understanding algebra, such as knowing how to find common factors, and there is never time in the class to catch her up to speed on these missing concepts. Years ago, her inclusive math classes stopped working on the time-telling and money skills she really needs to have in the real world, and which she still doesn't grasp well enough.

The decision as to whether or not to enroll in algebra should depend on the reason you and the student want him to attend the class. If your purpose is primarily social so he can be with his friends and work with them, it is different than if he is nearing the end of his schooling and is not ready to transition to work or to other postsecondary pursuits. If the student has been doing well in learning math survival skills, but needs more instruction, you might have him attend the algebra class and receive extra practice on math survival skills with a tutor or parent in a one-on-one situation.

Because the *Teaching Math* books are intended to cover basic and advanced survival skills, lessons in algebra will not be given in this book. However, if your student is going to participate in a general algebra class, it may help to know the "big ideas" that he should be able to understand. Possible topics that may have relevance to the student are:

1. Positive and negative numbers (as illustrated by a number line)
2. Solving equations using the principle of balance

3. Some scientific notation, including exponents and parentheses
4. Fractions to decimals, decimals to fractions
5. Writing and interpreting graphs and data (if covered in your school's algebra class)
6. Calculating perimeters and areas (if not covered in geometry)
7. Important algebra vocabulary

HANDS-ON ALGEBRA TEACHING RESOURCES

Just as with other areas of math learning, students with Down syndrome, autism, or other concrete learners have an easier time understanding algebra if they are allowed to use manipulatives. Although the student will probably have to follow the curriculum adopted by the school district, there are programs available that use manipulatives and hands-on activities that can help with comprehension. Programs available include:

1. Algebra Lab
2. Algeblocks
3. Algebra Tiles
4. Hands-on Equations

Algebra Lab. A commercial program called *Algebra Lab* (Creative Publications) can be used to supplement algebra textbooks or be used as a stand-alone program. The purpose of this program is to build students' understanding of key algebraic concepts through the use of Algebra Lab Gear, which are hands-on manipulatives. The program consists of colorfully illustrated student books, teachers' guides, and Lab Gear blocks in two colors—yellow for whole numbers and blue for variables such as x and y. The blocks are used with a corner piece that is sort of an expanded number line. The student books are well illustrated and full of useful applications and word problems that relate to most people's daily living.

Algeblocks. Another similar program is *Algeblocks* (ETA/Cuisenaire). This program also has three-dimensional blocks to represent variables. Algeblocks, however, comes with several work mats and a clear plastic factor track in the shape of a plus sign that can be used to represent equations, negative and positive numbers, and operations with unknown variables.

Algebra Tiles. Many teachers have already used *Algebra Tiles* (ETA/Cuisenaire) as a supplement to general education math materials, since they came on the market before Algebra Lab Gear and Algeblocks. Algebra Tiles, however, are two dimensional, and can be difficult for students to handle because they are so thin. They also do not have a corner piece or a factor track to provide the structure for the student's manipulations. On the plus side, they can easily be reproduced on transparencies for use on overhead projectors.

Hands-on Equations. The *Hands-on Equations* program, designed by Dr. Henry Borenson (Borenson and Associates), serves as a visual and kinesthetic teaching system for introducing some algebraic concepts, especially solving problems using equations. The equation is reproduced using a balance scale and small numbered cubes and colored pawns (that look like game tokens) to represent the unknown factor. The student simplifies the equation by doing something equally to each side of the balance until he finds the value of the pawn. Other cubes can be used to represent negative values. The program is not as comprehensive as Algebra Lab Gear or Algeblocks, but it is intended for students as young as third grade up to eighth or ninth grade.

Would Your Student Benefit from Algebra Class?

As you can see, I am not saying that students with Down syndrome or other hands-on learners can't take a general education algebra class. I do think that some adaptations and supplementations will have to be prepared so the student feels successful. Just as parents and teachers make sure that the student experiences and learns the "big ideas" of a general education class in social studies and science, the IEP team should write objectives that insure that the general education math teacher concentrates on the major ideas in algebra as far as the hands-on student is concerned.

Joe had been included successfully in general education math classes during his elementary and beginning middle school years. He was scheduled into a general education algebra class when he was in eighth grade. Within two weeks of starting the class, he indicated that he didn't like it and didn't want to go. His parents looked at the textbook and saw that the class work on equations was not much more difficult than the word problems he had already shown himself capable of doing.

Eventually, Joe blurted out that he didn't understand all the big words the teacher used in class. Joe pointed out some words he did not know in the textbook, but he was more upset about the unfamiliar words that the teacher was using. Words such as polynomials, commutative law, distributive property, equivalent, coefficient, and integer made him feel like he was hearing another language. Besides, some of the words that he understood were being used a different way in algebra class—for example, simplify (an equation), mean, power, squared, and function. Before Joe could be successful in that algebra class, he had to be taught some of the basic vocabulary used in algebra. It also might be helpful if the math teacher used simpler language or at least defined the unfamiliar word in the sentence in which he was using it. For example:

"For addition, you can use the commutative law, which says that you can add the numbers in any order."

When I was in elementary and secondary school, I always did well in math. However, as a girl of my generation, I was discouraged from taking math classes beyond algebra and geometry. I evidently knew enough math to get into college at that time. Much later, however, when I was working on my Ph.D., I had to take several college statistics courses. When I bought the textbook for the class, the first page read something like, "If it has been more than two years since you have taken college algebra, you may have difficulty with this statistics class." Well, it had been at least twenty-five years since I had taken algebra! And it turned out that statistics class was frustrating for me! I needed to be tutored by my high school-aged son when we got beyond mean, median, and mode.

Looking back at the class, the thing I remember most were the Greek letters that were used for many functions. The other students knew what they represented and used them easily. That "foreign" vocabulary made me feel like an outsider just pedaling as fast as I could to keep up with the rest of the class. I frequently tuned out the teacher's words in class and then took the book home where I could figure out the ideas myself—and get help from my family. I took two more statistics courses later and became very comfortable with the concepts; however, I have never forgotten my discomfort and frustration in that class trying to process words and concepts that were foreign to me, but not to anybody else.

I think of that situation in my statistics class when discussing placement in upper level math courses for students with cognitive disabilities. Again, in thinking through the decision, I think it is important to begin by clarifying what you or the student hope to accomplish by enrolling in algebra class:

- Is the goal for him to try to pass the regular algebra class? Perhaps because passing algebra is a requirement for high school graduation in your state or school district?
- Is the goal for him to participate in modified/adapted activities and learn just some important principles of algebra? Perhaps because he enjoys a challenge or has already mastered the "consumer" math he needs to be independent as an adult?
- Is the goal for the student to be "fully included" in all classes? Perhaps because the parents or IEP team thinks that other placements would be detrimental to the student's social skills or self-esteem? Or because they feel that it is the student's *right* to be completely included? Sometimes parents have fought so hard for their child's inclusion that they don't want to be perceived as taking a step backwards from full inclusion.

Once the IEP team is clear on the reason the student is taking algebra, goals must be written with an eye to this overall purpose. Regardless of the goals, the student will likely need support at home, pre-teaching and reinforcement of vocabulary and concepts, and perhaps tutoring outside of class with one of the hands-on programs described above. The teacher and parents should constantly check to determine whether the student needs help with vocabulary, with the big ideas, with the pace of the class, etc. Whatever the goals for the student, it is important that the student feel that he is making progress in the class so all his work is worth it.

Appendix A: Calculators

There are many different types of calculators that may be useful for students who are learning survival math skills. Originally I started with a standard calculator that was frequently used in schools: the *Calc-U-Vue.* Two of the young adults I worked with, however, were unable to accurately press the keys on that small calculator. We then switched to *large, inexpensive calculators,* which were a better match for their fine motor skills.

Later we began using a *talking calculator* (there are a variety of models available for under $20). The students with Down syndrome were having difficulty entering the numbers in the calculator accurately. When they heard the numbers spoken after they entered them, they were able to be more accurate. For students working in a classroom, a *printer calculator* can be helpful, as it enables the teacher to check both the students' ability to solve problems and their accuracy in entering data.

I've found that as students get more proficient with using the calculator, some of them can benefit from a calculator that displays the whole problem as it is inputted. For example, the calculator shows 4 + 5 = 9, not just each number separately. One calculator currently on the market that has this function is the *See 'N' Solve Visual Calculator.*

When working with money, hands-on learners often encounter a problem in using a calculator. That is, the calculator does not always finish the answer with 2 places to the right of the decimal point for cents (for example, it shows $4.5 instead of $4.50, or it

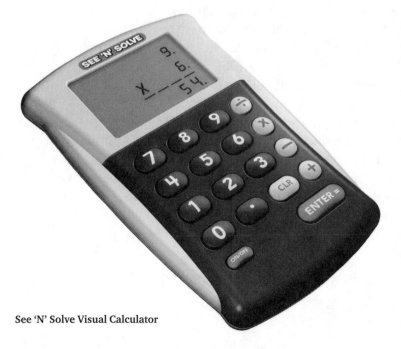

See 'N' Solve Visual Calculator

shows $3.33333 instead of $3.33. The ***Money Calc*** by PCI Educational Publishing ($20-25) has a feature that always answers in 2 places for cents. It also has a button that can be used to compute tips and tax. However, it is brightly colored and is probably not age-appropriate for secondary students.

An alternative for older students and adults is to use the ***Texas Instruments TI-15*** calculator because it has a feature that can be set to show only two places right of the decimal point. The TI-15 calculator (and the older TI-12) can also display fractions and remainders in division.

Texas Instruments TI-15.

Photo used with permission from TI. http://education.ti.com.

Tips on Use of the Calculator

1. Let the students play a little bit with the calculator when they first receive it. All of us like to fool around with a new toy. If you give them some time to experiment, they should be more able to use it seriously for work.
2. Have them enter the numbers slowly. Accuracy is much more important than speed.
3. Give plenty of practice using the calculator's keyboard. The numbers are arranged differently than the numbers on a phone. Especially if the students have a lot of experience dialing phones, they will need practice to train their motor skills to locate the numbers on a different keyboard.
4. Use the calculator to verify problems you have worked out with manipulatives. At other times, use the manipulatives to verify the answer you have received from the calculator.
5. Get in a rhythm when using the calculator. You can chant the sequence:
   Number (whatever the number is) as you push the number.
   Look (check the number in the window)
   Plus (or whatever the operation sign is)
   Number (whatever the second number is)
   Look
   Equals (as you press the = sign) to get the answer
6. Consider having the use of a calculator put in the student's IEP. Students may be allowed to use calculators on state assessments on problems that do not involve the knowledge of the math facts. For example, if word problems are being assessed, a calculator may be permitted. Of course, when the purpose is to assess math facts, calculators are not allowed on tests.

Appendix B: Teaching Materials

- B-1: Earn & Pay Game Board
- B-2: Earn & Pay Game Cards
- B-3: Money Total Slips
- B-4: Play Currency
- B-5: Journey Game Board
- B-6: Journey Game Problem Cards
- B-7: Number Line
- B-8: Number Line Addition
- B-9: Number Line Activities
- B-10 Addition Worksheet
- B-11: Menu Plan Form
- B-12: Calories and Food
- B-13: Addition Word Problem Form
- B-14: Addition Word Problems Worksheet
- B-15: Cookie Worksheet
- B-16: Take-Away Subtraction Worksheet
- B-17: Number Line Subtraction
- B-18: Types of Subtraction Worksheet
- B-19: Subtraction with Larger Numbers Worksheet
- B-20: Word Problem Form (Subtraction and Addition)
- B-21: Subtraction Worksheet (Using Structured Form)
- B-22: Mixed Addition and Subtraction Word Problem Worksheets
- B-23: Menu Board
- B-24: Number Cube Pattern
- B-25: Sixes Grid Game
- B-26: Turnarounds
- B-27: Multiplying by Ten Worksheet
- B-28: Calculator Multiplication Worksheet
- B-29: Batter Up Game Board
- B-30: Batter Up Game Cards (and Answers)

- B-31: Times 9 Chart
- B-32: 2s and 3s Multiplication Fact Cards
- B-33: 4s Multiplication Fact Cards
- B-34: Times Cover Game
- B-35: Multiplication Tables—6s to 9s
- B-36: Multiplication by One Digit
- B-37: Multiplication Facts Checkers
- B-38: Blank Checkerboard
- B-39: Word Problem Form (All Operations)
- B-40: Addition & Multiplication Story Problems
- B-41: Simple Calculator Division Problems
- B-42: Division Tic-Tac-Toe Game
- B-43: Calculator Division Problems
- B-44: Division Problems Checked by Multiplication
- B-45: Fraction Strips
- B-46: Domino Fraction Boards
- B-47: Domino Fraction Cards
- B-48: Fraction Strips Worksheet
- B-49: Equivalent Fractions Worksheet
- B-50: Perimeter Worksheet
- B-51: Andy's Play Area
- B-52: Square Inch Pieces
- B-53: Perimeter and Area Worksheet
- B-54: Quarter Clock
- B-55: Activity Strips for Time Matching Game
- B-56: Game Cards for Time Matching Game
- B-57: Answer Sheet for Time Matching Game
- B-58: Food Temperature Danger Zone Chart
- B-59: Place Value Money Chart
- B-60: Reading Prices Worksheet
- B-61: Blank Checks
- B-62: Calculator Money Worksheet
- B-63: Next-Highest-Dollar Worksheet
- B-64: Large Coin Pictures
- B-65: Small Coin Pictures (nickels, dimes)
- B-66: Nickels Worksheet
- B-67: Dimes Worksheet
- B-68: Dimes and Nickels Worksheet
- B-69: Quarters Worksheet
- B-70: Mixed Coin Worksheet
- B-71: Mixed Coin (Including Pennies) Worksheet
- B-72: Coin Bingo Cards
- B-73: Totaling Your Shopping List
- B-74: Shopping for One Worksheet
- B-75: Which Price Is Best? Game Board
- B-76: Which Price Is Best? Game Cards
- B-77: Which Price Is Best? Answer Cards

Get a cold.
Lose one turn.

PAY $

Go to a game.

EARN $

Mow the lawn.

Babysit.

PAY $

Go to an amusement park.

EARN

EARN $

Get birthday gift of money.

FINISH

START

PAY $

Buy toys or items for collection.

PAY $

Go bowling.

PAY $

Buy

(Cut and join with other half.)

EARN $

Buy a music CD.

PAY $

Sell stuff at a garage sale.

EARN $

Missing the bus.
Go back 2 spaces.

&PAY

EARN $

Sell lemonade.

PAY $

Go to the movies.

pizza.

EARN $

Get a prize for
most cookie sales.

EARN $

Clean the basement.

SOAP

Did a good deed.
Go ahead 2 spaces.

Earn & Pay Game Cards

PAY $ $ 10	PAY $ $ 15	PAY $ $ 25
PAY $ $ 13	PAY $ $ 26	PAY $ $ 35
PAY $ $ 40	PAY $ $ 30	PAY $ $ 8

Earn & Pay Game Cards

PAY $ $ 15	PAY $ $ 10	PAY $ $ 21
PAY $ $ 20	PAY $ $ 10	PAY $ $ 11
PAY $ $ 8	PAY $ $ 20	PAY $ $ 22

Earn & Pay Game Cards

EARN $	EARN $	EARN $
$ 10	$ 15	$ 20

EARN $	EARN $	EARN $
$ 13	$ 16	$ 30

EARN $	EARN $	EARN $
$ 21	$ 18	$ 8

Earn & Pay Game Cards

EARN $	EARN $	EARN $
$ 12	$ 10	$ 14
EARN $	EARN $	EARN $
$ 20	$ 7	$ 11
EARN $	EARN $	EARN $
$ 8	$ 20	$ 9

Earn & Pay Game Cards

PAY $	PAY $	PAY $
$ 10	$ 5	$ 8
PAY $	PAY $	PAY $
$ 11	$ 2	$ 7
PAY $	PAY $	PAY $
$ 9	$ 10	$ 8

Earn & Pay Game Cards

PAY $ **$ 4**	PAY $ **$ 10**	PAY $ **$ 11**
PAY $ **$ 7**	PAY $ **$ 9**	PAY $ **$ 11**
PAY $ **$ 8**	PAY $ **$ 4**	PAY $ **$ 3**

Earn & Pay Game Cards

EARN $ $10	EARN $ $5	EARN $ $2
EARN $ $4	EARN $ $6	EARN $ $7
EARN $ $9	EARN $ $3	EARN $ $6

Earn & Pay Game Cards

EARN $ $ 6	EARN $ $ 10	EARN $ $ 5
EARN $ $ 4	EARN $ $ 7	EARN $ $ 11
EARN $ $ 3	EARN $ $ 2	EARN $ $ 9

MONEY TOTAL SLIP

Name:

Name of bill	Number of bills	Total amount of money in this bill
$1.00		
$5.00		
$10.00		
$20.00		
$50.00		
	Final Total:	

MONEY TOTAL SLIP

Name:

Name of bill	Number of bills	Total amount of money in this bill
$1.00		
$5.00		
$10.00		
$20.00		
$50.00		
	Final Total:	

MONEY TOTAL SLIP

Name:

Name of bill	Number of bills	Total amount of money in this bill
$1.00		
$5.00		
$10.00		
$20.00		
$50.00		
	Final Total:	

MONEY TOTAL SLIP

Name:

Name of bill	Number of bills	Total amount of money in this bill
$1.00		
$5.00		
$10.00		
$20.00		
$50.00		
	Final Total:	

MONEY TOTAL SLIP

Name:

Name of bill	Number of bills	Total amount of money in this bill
$1.00		
$5.00		
$10.00		
$20.00		
$50.00		
	Final Total:	

MONEY TOTAL SLIP

Name:

Name of bill	Number of bills	Total amount of money in this bill
$1.00		
$5.00		
$10.00		
$20.00		
$50.00		
	Final Total:	

PLAY CURRENCY $1

PLAY CURRENCY $5

PLAY CURRENCY $10

PLAY CURRENCY $20

PLAY CURRENCY $50

Journey Game Problem Cards
(Multiplication—Numbers, 1 and 2 digits)

YM1

$$42 \times 6 = \underline{\hspace{2cm}}$$

YM2

$$5 \times 54 = \underline{\hspace{2cm}}$$

YM3

$$7 \times 22 = \underline{\hspace{2cm}}$$

YM4

$$5 \times 70 = \underline{\hspace{2cm}}$$

YM5

$$8 \times 9 = \underline{\hspace{2cm}}$$

YM6

$$5 \times 7 = \underline{\hspace{2cm}}$$

YM7

$$7 \times 43 = \underline{\hspace{2cm}}$$

YM8

$$6 \times 22 = \underline{\hspace{2cm}}$$

Journey Game Problem Cards
(Division—Numbers, 1 and 2 digits)

ED1 $72 \div 8 = \underline{\hspace{2cm}}$	ED2 $24 \div 6 = \underline{\hspace{2cm}}$
ED3 $60 \div 6 = \underline{\hspace{2cm}}$	ED4 $36 \div 6 = \underline{\hspace{2cm}}$
ED5 $81 \div 9 = \underline{\hspace{2cm}}$	ED6 $35 \div 7 = \underline{\hspace{2cm}}$
ED7 $45 \div 5 = \underline{\hspace{2cm}}$	ED8 $21 \div 7 = \underline{\hspace{2cm}}$

Journey Game Problem Cards
(Subtraction—Numbers, 1 and 2 digits)

OS1 $42 - 6 = $ ____	OS2 $85 - 54 = $ ____
OS3 $27 - 22 = $ ____	OS4 $95 - 70 = $ ____
OS5 $8 - 3 = $ ____	OS6 $25 - 7 = $ ____
OS7 $7 - 4 = $ ____	OS8 $36 - 22 = $ ____

Journey Game Problem Cards
(Addition—Numbers, 1 and 2 digits)

TA1 $6 + 18 = \underline{\hspace{2cm}}$	TA2 $17 + 9 = \underline{\hspace{2cm}}$
TA3 $6 + 7 = \underline{\hspace{2cm}}$	TA4 $8 + 7 = \underline{\hspace{2cm}}$
TA5 $28 + 9 = \underline{\hspace{2cm}}$	TA6 $5 + 7 = \underline{\hspace{2cm}}$
TA7 $7 + 15 = \underline{\hspace{2cm}}$	TA8 $41 + 7 = \underline{\hspace{2cm}}$

Journey Game Problem Cards
(Addition—Word problems, 1-3 digits)

WA1 Michael read 27 books last year. This year he read 32 books. How many books has he read altogether?	**WA2** Brad has 25 stamps in his collection. He got 22 new stamps for his birthday. How many stamps does he have now?
WA3 Heather drives her daughter to school and picks up her friend Katie on the way. If she drives 2 miles to Katie's house and then 3 miles more to the school, how many miles has she driven to get the girls to school?	**WA4** Jeff and his family are eating spaghetti. Jeff has 3 meatballs on his spaghetti, Heidi has 2 meatballs, Katri has 3 meatballs, and Kyle has 5 meatballs. How many meatballs does the family have altogether?
WA5 Mark is on the football team that plays a 9 game season. There are 4 away games where the team is out of town. How many home (at their own stadium) games does he play?	**WA6** Jake went to the zoo. He saw 7 adult tigers and 2 baby tigers. How many tigers did he see at the zoo?
WA7 The cheerleading squad was collecting canned goods for the flood relief in the south of the state. Carol brought 9 cans, Marie 11 cans, Allison 23 cans, and Ellen 22 cans. How many cans did they get?	**WA8** McKenzie and her brother loved to play video games. Her brother's highest score was 203. She beat her brother's score by 10 points. What was her score?

Journey Game Problem Cards
(Subtraction—Word problems, 1-3 digits)

WS1

Veronica has a paper route. She usually delivers 37 papers each morning. She received notice that 4 families on her route were moving away. How many newspapers does she have to deliver now?

WS2

Shelby is cooking the pasta for dinner. The recipe says to boil the pasta for 9 minutes. The pan has been boiling for 5 minutes already. How many minutes more does she have to let it boil?

WS3

Ryan is allowed to use the computer for 45 minutes to e-mail his friends. He has been on the computer for 15 minutes. How much more time does he have left to use the computer?

WS4

Paul is saving up to buy a $200 guitar. He has saved $125 already. How much more does he have to save?

WS5

The school bus has 32 seats on it. The bus has picked up 20 children. How many empty seats are there on the bus?

WS6

Kayli has a regular allowance of $5.00 a week. This week she has to repay her brother Brad the $1.25 that she borrowed. How much money will she have to spend this week?

WS7

The discount store is having a sale on DVD's. All DVD's are $16 each. Before the sale, all the DVD's were $20 each. How much money will Ann save if she buys one DVD?

WS8

Jerry found 19 ants around the kitchen sink on Sunday. On Monday he saw 30 ants. How many more ants did he see on Monday than on Sunday? (He had better get the ant spray!)

Journey Game Problem Cards
(Multiplication—Word problems, 1 and 2 digits)

WM1

Mark and his 2 friends each bought eight bananas. How many bananas did they buy in all?

WM2

Rowan put 6 stickers on each row of a sticker album. If there were 7 rows on a page, how many stickers are on the whole page?

WM3

Jake has 3 toy boats. His 2 friends also have 3 boats each. How many boats are there all together? *(Two steps)*

WM4

Eric has 3 toy trains in the basement. If each train has 10 cars (including the engine and caboose) each, how many train cars does he have all together?

WM5

Season collects play frogs. She puts 3 frogs on each windowsill. If her house has 11 windows, how many frogs does she have in the windows all together?

WM6

Ryan's schoolbus has 42 seats. How many students can ride on 4 schoolbuses?

WM7

How many eggs are in 4 cartons? Hint—each carton holds 12 eggs.

WM8

Paul took 6 pictures of Hawaii every day he was on vacation. If he spent 5 days in Hawaii, how many pictures did he take of Hawaii?

Journey Game Problem Cards
(Division—Word problems, 1 and 2 digits)

WD1

Katri has 24 chocolate kisses. If she divides them equally between her 6 friends, how many will each friend get? (Do not include Katri; she already ate her share.)

WD2

The stamp club ordered a pizza for their meeting. The pizza cost $12. If the 4 members of the club shared the cost equally, how much should each member pay?

WD3

Allison dealt out 40 cards for the Money game. If there were 4 card players, how many cards did each player get?

WD4

Lindsay dug up 18 plants from her garden. She wanted to give her 3 friends an equal number of plants. How many did she give to each friend?

WD5

Brian went caroling at Christmas time. He sang a total of 24 songs. If he sang the same number of carols at each of 6 houses, how many carols did he sing at each house?

WD6

Scott liked tee shirts with wild animal pictures on them; he had 12 of them. If his mom and dad each got him the same number of tee shirts, how many did each one of them get?

WD7

Steve made scrambled eggs for the whole family (4 members) and used 12 eggs. If they got the same amount of eggs, how many eggs did each get?

WD8

Scott's friends asked him to reset their watches to Daylight Savings Time. Each friend had him fix 2 watches and he fixed eight watches total. How many friends had Scott fix their watches?

Journey Game Problem Card Anwers

NUMBER PROBLEMS	WORD PROBLEMS
YM1 = 252	WA1 = 59 books
YM2 = 270	WA2 = 47 stamps
YM3 = 154	WA3 = 5 miles
YM4 = 350	WA4 = 13 meatballs
YM5 = 72	WA5 = 5 games
YM6 = 35	WA6 = 9 tigers
YM7 = 301	WA7 = 65 cans
YM8 = 132	WA8 = 213 points
ED1 = 9	WS1 = 33 newspapers
ED2 = 4	WS2 = 4 minutes
ED3 = 10	WS3 = 30 minutes
ED4 = 6	WS4 = $75
ED5 = 9	WS5 = 12 seats
ED6 = 5	WS6 = $3.75
ED7 = 9	WS7 = $4.00
ED8 = 3	WS8 = 11 ants
OS1 = 36	WM1 = 24 bananas
OS2 = 31	WM2 = 42 stickers
OS3 = 5	WM3 = 9 boats
OS4 = 25	WM4 = 30 cars
OS5 = 5	WM5 = 33 frogs
OS6 = 18	WM6 = 168 students
OS7 = 3	WM7 = 48 eggs
OS8 = 14	WM8 = 30 pictures
TA1 = 24	WD1 = 4 chocolate kisses each
TA2 = 26	WD2 = $3.00
TA3 = 13	WD3 = 10 cards
TA4 = 15	WD4 = 6 plants
TA5 = 37	WD5 = 4 carols
TA6 = 12	WD6 = 6 tee shirts each
TA7 = 22	WD7 = 3 eggs
TA8 = 48	WD8 = 4 people

Number Line (from Hundreds Chart)

(If possible, laminate the chart.) Cut the chart so that you have 10 lines with 10 numbers each. On the blank square next to the 10s, put a piece of Velcro or a rolled up piece of tape; attach to the back of the 1-91 column. You'll then be able to line up the numbers from 1-100 in a straight horizontal line!

1	2	3	4	5	6	7	8	9	10	
11	12	13	14	15	16	17	18	19	20	
21	22	23	24	25	26	27	28	29	30	
31	32	33	34	35	36	37	38	39	40	
41	42	43	44	45	46	47	48	49	50	
51	52	53	54	55	56	57	58	59	60	
61	62	63	64	65	66	67	68	69	70	
71	72	73	74	75	76	77	78	79	80	
81	82	83	84	85	86	87	88	89	90	
91	92	93	94	95	96	97	98	99	100	

Number Line Addition

1) $11 + 7 =$ ___

2) $12 + 0 =$ ___

3) $4 + 9 =$ ___

4) $3 + 1 =$ ___

5) $8 + 1 =$ ___

6) $8 + 8 =$ ___

7) $2 + 0 =$ ___

8) $4 + 12 =$ ___

9) $6 + 5 =$ ___

10) $5 + 12 =$ ___

Number Line Activities

The objective of frequently using a number line is to help the student develop a visual picture of the number line in his or her mind, and be able to relate the abstract numbers to it. Some activities for practicing using a number line are:

- Using a line 1–10, put the corresponding number of objects on each of the squares of the number line. For example, put 4 poker chips or buttons on the 4 of the number line.

- Using a line 1–30, count forward from any number except for 1. For example, count to 15, starting with 5.

- Using a line 1–30, name the numbers that are one more or one less than any number. For instance, point to 5, and say, "What number is one more than 5? What number is one less than 5?"

- Using a line 1–50, show which number is larger or smaller. For example, put a finger on number 31, then put another finger on 44. Which one is larger or closest to the end of the line (50)?

- Practice adding using the number line. For example, 8 plus 4 (hops) more (9,10,11,12) = 12.

- Practice subtracting using the number line. For example, 22 minus 5 (hops) less (21, 20,19,18,17) = 17.

- Using a line 1–50, skip count by 2s—hopping 2, 4, 6, 8, 10, 12, 14, etc.

- Using a line 1–50, skip count by 5s and by 10s.

- Skip counting (visualizing a number line) may be used for beginning multiplication and division. For example, if you start at 0 and take 5 hops, you land on 10 (5 x 2 = 10).

Addition Worksheet

1)
$$\begin{array}{r} 5 \\ +2 \\ \hline \end{array}$$

2)
$$\begin{array}{r} 99 \\ +90 \\ \hline \end{array}$$

3)
$$\begin{array}{r} 85 \\ +99 \\ \hline \end{array}$$

4)
$$\begin{array}{r} 95 \\ +11 \\ \hline \end{array}$$

5)
$$\begin{array}{r} 11 \\ +35 \\ \hline \end{array}$$

6)
$$\begin{array}{r} 39 \\ +22 \\ \hline \end{array}$$

7)
$$\begin{array}{r} 88 \\ +19 \\ \hline \end{array}$$

8)
$$\begin{array}{r} 56 \\ +76 \\ \hline \end{array}$$

9)
$$\begin{array}{r} 15 \\ +87 \\ \hline \end{array}$$

10)
$$\begin{array}{r} 41 \\ +40 \\ \hline \end{array}$$

11)
$$\begin{array}{r} 25 \\ +70 \\ \hline \end{array}$$

12)
$$\begin{array}{r} 70 \\ +26 \\ \hline \end{array}$$

Menu Plan Form

DAY 1		DAY 2		DAY 3	
FOOD	**CALORIES**	**FOOD**	**CALORIES**	**FOOD**	**CALORIES**
Main dish		Main dish		Main dish	
Fruit Vegetables		Fruit Vegetables		Fruit Vegetables	
Dessert		Dessert		Dessert	
Side dishes		Side dishes		Side dishes	
Condiments		Condiments		Condiments	
Other		Other		Other	
TOTAL:		**TOTAL:**		**TOTAL:**	

Calories and Food

FOOD	CALORIES
MAIN DISHES	
Hamburger	350
Big hamburger (Whopper)	660
Burrito	400
Chicken sandwich	530
Roast beef andwich	450
Lunch meat sandwich	350
Ham & cheese sandwich	560
Pizza (2 pieces)	450
Big Mac	570

FOOD	CALORIES
FRUITS / VEGETABLES	
Lettuce	10
Green beans	10
Mushrooms	10
Tomato	10
Corn	80
Fruit cocktail (1/2 cup)	160
Peaches (1/2 cup)	200
Apple	85
Banana	129
Orange	100
Strawberries	50

DESSERTS	
Cherry pie	400
Shake	450
Ice cream	350
Frozen yogurt	180
Sundae	750
Candy bar	250
Choc. chip cookie	100
Cake	300
Jell-O	100

DRINK	
Diet pop	5
Regular pop	160
Milk	100
Coffee/Tea	5
Coffee/Tea (cream & sugar)	100
Water	0

SIDE DISHES	
French fries	300
Potato chips	300
Bread (1 slice)	100
Baked potato	100
Mashed potatoes	150
Rice	180
Macaroni & Cheese	280
Baked beans	150

CONDIMENTS	
1 T. butter	100
1 T. Margarine	100
Gravy (1/4 cup)	100
Mayonnaise	100
Ketchup	16
Mustard	16
Relish	20

Check the items wanted for your meal. Write the calorie count for each selection; add to find the total calories.

Addition Word Problem Form

1) READ and understand the problem. (Write needed numbers with their labels, e.g., 4 desks) (Write the answer sentence with a blank.)	**2) DRAW, use number line, or act out problem.**
3) FIND the operation. Will the answer be larger than the biggest number in the problem?	**4) CIRCLE the operation.** **Larger = Addition**

5) WRITE THE NUMBER SENTENCE.
 	
        = _____ .
 	
    Solve the problem =
	6) WRITE ANSWER in the blank of the sentence in square 1.

Addition Word Problems Worksheet

1. Ellen collects small pieces of doll furniture. She has 2 tables, 2 sofas, 5 beds, 3 lamps, and 6 chairs. How many pieces of doll furniture does she have?

2. Kristy is bringing pop for the school party. She brought 1 carton of orange pop, 2 cartons of cola, and 4 cartons of lemon-lime pop. How many cartons did she bring?

3. Greg drove his truck 213 miles on Monday, 40 on Tuesday, 111 miles on Wednesday, and 78 miles on Friday. How many miles did he drive in that week?

4. Carol played 4 video games before dinner. After dinner she played 4 more video games. After she went to bed, she sneaked out and played 4 more video games. How many video games did she play that day?

5. Cindy and Nancy went out to eat for lunch. They both ordered French fries. Cindy got 23 fries and Nancy got 31. Now is that fair? How many French fries did they get altogether?

6. Linda collects different pencils. She has 4 pencils with the Olympics logo stamped on them, 3 of them were signed by NASCAR drivers, 4 have Disney characters on them, and 3 have feathers on them. How many pencils does she have?

Cookie Worksheet

Photocopy and cut out 12 cookies for the instructor and for each student.

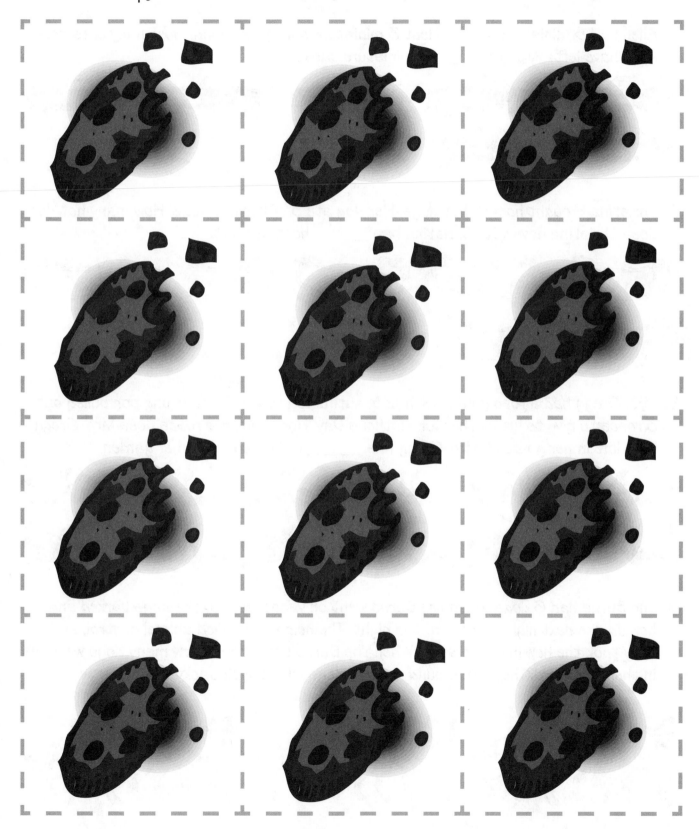

Take-Away Subtraction Worksheet

Cross out the items that have been taken away. Write down the number of items left.

1. Allen had 8 miniature cars. He lent 3 miniature cars to a friend. How many cars does Allen have left? Allen has _____ miniature cars left.

2. Christina brought home 7 hot dogs. Her dog ate 3 of the hot dogs. How many hot dogs does Christina have left? Christina has _____ hot dogs left.

3. Mrs. Gregg had a rose garden with 12 beautiful red roses. Alex's young son pulled out 3 roses to give to his mother for Mother's Day. How many red roses does Mrs. Gregg have left in her garden? Mrs. Gregg has _____ red roses left in her garden.

4. The Burns had 6 dogs. When the Burns went on vacation, 4 of the dogs barked and howled the next night—most of the night. The neighbors called animal control, and they took the howling dogs away. When the Burns came come, how many dogs were left in their backyard? The Burns found _____ dogs left in their backyard.

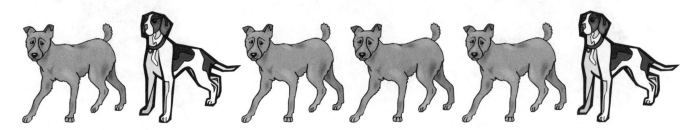

Number Line Subtraction

1) $9 - 4 = \underline{\quad}$

2) $15 - 6 = \underline{\quad}$

3) $26 - 3 = \underline{\quad}$

4) $7 - 6 = \underline{\quad}$

5) $30 - 5 = \underline{\quad}$

6) $18 - 1 = \underline{\quad}$

7) $21 - 2 = \underline{\quad}$

8) $10 - 7 = \underline{\quad}$

9) $25 - 4 = \underline{\quad}$

10) $16 - 8 = \underline{\quad}$

Types of Subtraction Worksheet

Write the type of subtraction on the line after the problem. You may just write the letters as shown. Choose from:

T—TAKE-AWAY　**C**—COMPARISON　**HC**—HOW MUCH CHANGE　**HM**—HOW-MUCH-MORE

1. Heather is saving to buy a CD. The CD costs $8, but she has only $5 saved up. How much more money does she need?

   Type of subtraction_____

2. Mark wanted to buy a bike helmet. The helmet cost $24. Mark gave the cashier $30. How much change should he get back?

   Type of subtraction_____

3. Paul's cat had 6 new kittens, and Ryan's dog had 4 new puppies. Who has the most new animals? How many more?

   Type of subtraction_____

4. You have 7 packs of gum. Your brother takes 3 packs. How many do you have left?

   Type of subtraction_____

5. Karryl's family had their picture taken. The photographer charged $12 for the pictures. Karryl gave him $20. Does she get any change back? How much?

   Type of subtraction_____

6. Jeff has 10 action figures (Power Rangers). He brought 6 of them in his suitcase when he came to visit his grandma. How many action figures are still left at home?

   Type of subtraction_____

Subtraction with Larger Numbers

1)
$$935 - 0$$

2)
$$61 - 6$$

3)
$$14 - 0$$

4)
$$49 - 35$$

5)
$$148 - 96$$

6)
$$45 - 27$$

7)
$$137 - 12$$

8)
$$21 - 16$$

9)
$$980 - 307$$

10)
$$12 - 1$$

11)
$$889 - 555$$

12)
$$706 - 701$$

Word Problem Form (Subtraction and Addition)

1) READ and understand the problem. (Write needed numbers with their labels, e.g., 4 desks) (Write the answer sentence with a blank; e.g., Tom has ____ books left.)	**2) DRAW, use number line, or act out problem.**
3) FIND the operation. Will the answer be larger than the biggest number in the problem?	**4) CIRCLE the operation.** <div align="center">Larger = Addition Smaller = Subtraction</div>
5) WRITE THE NUMBER SENTENCE. = _____ . Solve the problem =	**6) WRITE ANSWER in the blank of the sentence in square 1.**

Subtraction Worksheet (Using Structured Form)

Read the following problems carefully. Use the Word Problem Form (Addition and Subtraction) from Appendix B for each problem.

1. Jared works at the zoo. He needs to build a 97 foot fence around the bear area. He has built 62 feet already. How many feet of fencing does he have left to build? Jared has _____ feet left to build.

2. There were 18 elephants drinking at the river. Five elephants walked away. How many elephants are left drinking at the river? There are _____ elephants left drinking at the pool.

3. Shawn bought a dozen (12) eggs. He used 6 eggs to make French toast. How many eggs does he have left? Shawn has _____ eggs left.

4. Leslie watches 19 hours of TV a week. Her mother wants her to cut back to watching 12 hours a week. How many fewer hours of TV can she watch? Leslie can watch _____ fewer hours of TV a week.

5. Ann made 16 home runs in softball this year. Last year she hit only 5 home runs. How many more home runs did she make this year? Ann made _____ more home runs this year than last year.

Mixed Addition & Subtraction Word Problem Worksheets

Use the word problem form (Appendix B) to solve these word problems. Some problems are addition problems, and some are subtraction problems.

1. There were 40 people rollerblading at the skating park. 25 more people came to the park. How many people were now rollerblading at the park? There were _____ people rollerblading in all.

2. Barbara sold 9 boxes of candy for a school contest. Her goal was to sell 15 boxes so she could get a prize. How many more does she need to sell to get the prize? Barbara needs to sell _____ more boxes of candy.

3. Nick's father has a rope and a hose in the garage. The rope is 58 inches long, and the hose is 21 inches long. How much longer is the rope than the hose? The rope is _____ inches longer than the hose.

4. In a theater with 100 seats, 80 seats were full. 15 more people came. How many people are there in the theater now? There are _____ people at the theater now.

5. Greg made 23 baskets at the basketball game. He also missed 10 shots. How many shots did he take in all? Greg took _____ shots at the basketball game.

6. Emily had 25 spelling words for this week. On Friday she missed 3 of those spelling words. How many words did she spell correctly? Emily got _____ spelling words correct.

7. Ben got $5 in allowance this week. He also earned $7 babysitting. How much money has Ben received and earned this week? Ben has made _____ dollars this week.

8. There are 15 boys and 11 girls in Amber's class. How many students does Amber have in her class? Amber has _____ students in her class.

Menu Board

Popcorn
$1.00

Hot Dog
$2.00

Lemonade
$1.00

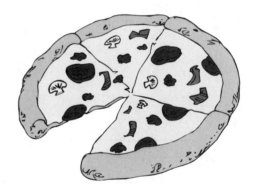

Pizza
$3.00

Candy
$2.00

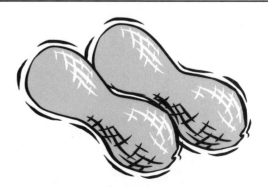

Peanuts
$1.00

Number Cube Pattern

Cut out pattern on solid lines.
Fold on dotted lines.
Glue opposite tabs together.

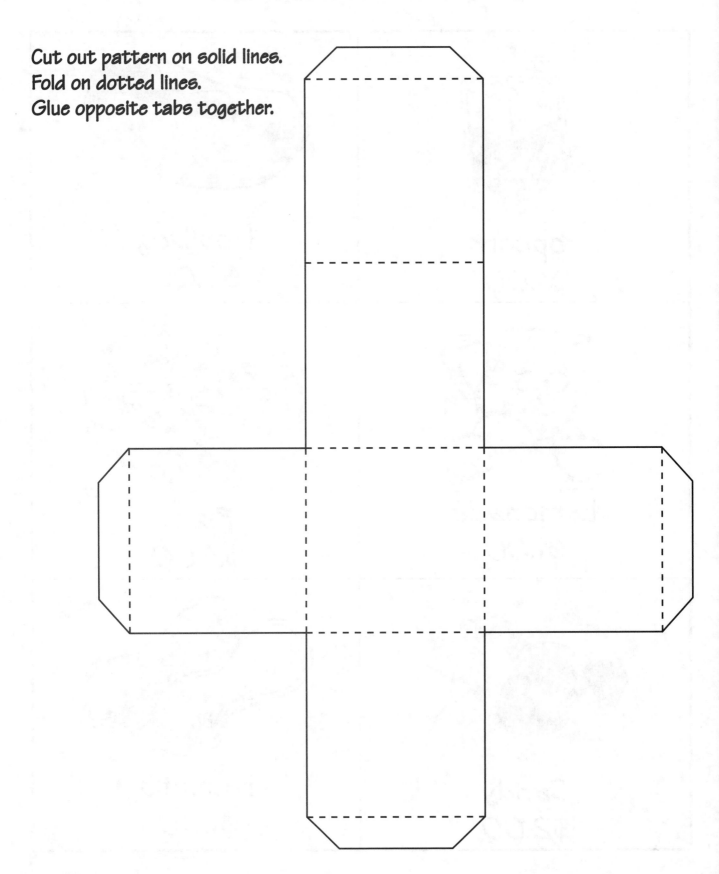

Sixes Grid Game

1	2	3	4	5	6
2					
3					
4					
5					
6					

INSTRUCTIONS

1. The first student rolls the dice. He multiplies the two numbers on the dice times each other (can use calculator). He locates the square where the two numbers intersect on the grid and writes the answer in that square.

2. The second student then throws the dice and fills in the answer on his grid.

3. Play continues until one person completely fills a line—across, down, or diagonally.

4. Later you may want to have them cover the entire card.

1	2	3	4	5	6
2					
3					
4					
5					
6					

Turnarounds

Problem	Turnaround
EXAMPLE: 2 x 3 = __6__	3 x 2 = __6__
4 x 2 = ____	
5 x 2 = ____	
5 x 6 = ____	
3 x 3 = ____	
1 x 2 = ____	
1 x 5 = ____	
6 x 5 = ____	
4 x 3 = ____	
5 x 3 = ____	
4 x 4 = ____	
4 x 5 = ____	
6 x 4 = ____	
3 x 1 = ____	
4 x 3 = ____	
5 x 3 = ____	
2 x 6 = ____	
2 x 5 = ____	
3 x 6 = ____	
6 x 4 = ____	

Multiplying by Ten Worksheet

Fold this paper into halves, fourths, and then eighths so that you have only one problem showing at a time.

1)

$5 \times 10 = \underline{\quad}$

2)

$3 \times 10 = \underline{\quad}$

3)

$2 \times 10 = \underline{\quad}$

4)

$4 \times 10 = \underline{\quad}$

5)

$6 \times 10 = \underline{\quad}$

6)

$7 \times 10 = \underline{\quad}$

7)

$4 \times 10 = \underline{\quad}$

8)

$8 \times 10 = \underline{\quad}$

Calculator Multiplication Worksheet

1)
$$76 \\ \times\,0$$

2)
$$41 \\ \times\,0$$

3)
$$56 \\ \times\,1$$

4)
$$38 \\ \times\,5$$

5)
$$32 \\ \times\,4$$

6)
$$62 \\ \times\,3$$

7)
$$61 \\ \times\,1$$

8)
$$67 \\ \times\,1$$

9)
$$66 \\ \times\,7$$

10)
$$47 \\ \times\,9$$

11)
$$20 \\ \times\,2$$

12)
$$33 \\ \times\,0$$

Calculator Multiplication Worksheet

1) 380
    × 0

2) 32
    × 6

3) 980
    × 4

4) 964
    × 4

5) 54
    × 7

6) 12
    × 6

7) 455
    × 1

8) 880
    × 9

9) 131
    × 1

10) 576
    × 7

11) 933
    × 6

12) 550
    × 2

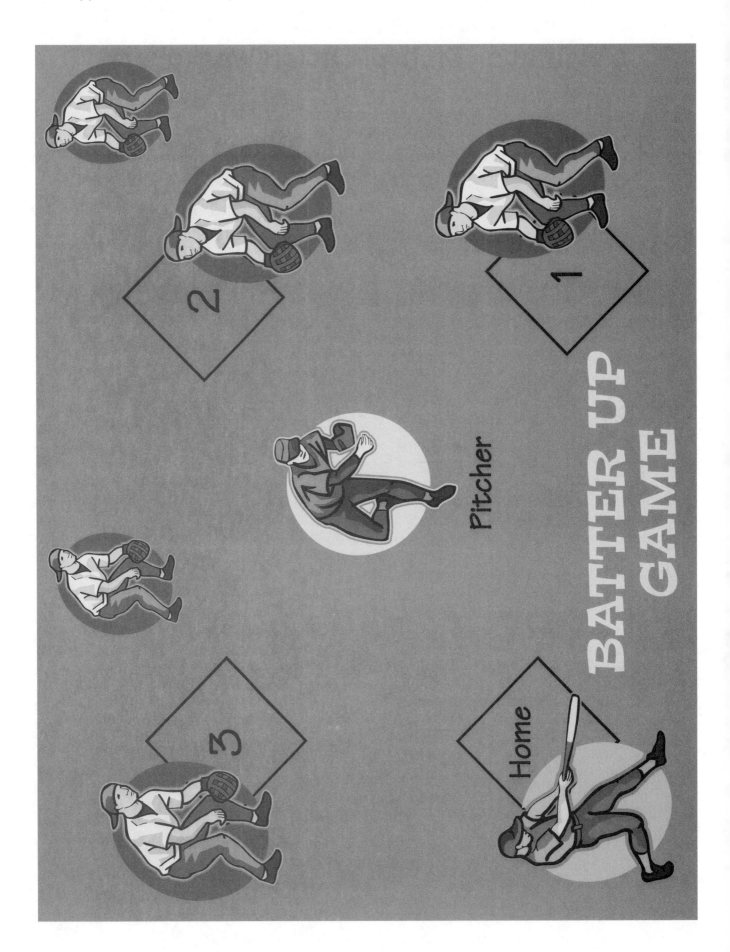

Batter Up Game Cards

Pitched ball hits you in the head. Walk to the first or next base.

Foul ball. Lose a turn.

Bunt. Move to first or next base.

Fielder caught your fly ball. You are out.

Fielder caught your fly ball. You are out.

Steal a base. Go ahead one base.

Steal a base. Go ahead one base.

You try to steal a base and are tagged out.

Steal a base. Go ahead one base.

Steal two bases.

Batter Up Game Cards

A 3 × 9 = or 12 × 33 =	A 6 × 7 = or 23 × 11 =
A 21 × 7 = or 8 × 9 =	A 48 × 7 = or 5 × 5 =
A 9 × 9 = or 45 × 22 =	A 7 × 7 = or 37 × 11 =
A 82 × 12 = or 4 × 9 =	A 5 × 9 = or 73 × 41 =
A 45 × 8 = or 8 × 3 =	A 22 × 5 = or 6 × 6 =

Batter Up Game Cards

B 7 x 5 = or 22 x 43 =	B 4 x 7 = or 56 x 11 =
B 21 x 3 = or 8 x 3 =	B 8 x 7 = or 31 x 22 =
B 7 x 9 = or 61 x 12 =	B 7 x 7 = or 37 x 11 =
B 32 x 42 = or 4 x 6 =	B 54 x 9 = or 8 x 8 =
B 22 x 4 = or 6 x 8 =	B 51 x 2 = or 8 x 4 =

Batter Up Game Card Answers

A 3 x 9 = 27 or 12 x 33 = 396	A 6 x 7 = 42 or 23 x 11 = 253
A 21 x 7 = 147 or 8 x 9 = 72	A 48 x 7 = 336 or 5 x 5 = 25
A 9 x 9 = 81 or 45 x 22 = 990	A 7 x 7 = 49 or 37 x 11 = 407
A 82 x 12 = 984 or 4 x 9 = 36	A 5 x 9 = 45 or 73 x 41 = 2993
A 45 x 8 = 360 or 8 x 3 = 24	A 22 x 5 = 110 or 6 x 6 = 36

Batter Up Game Card Answers

B 7 x 5 = 35 or 22 x 43 = 946	B 4 x 7 = 28 or 56 x 11 = 616
B 21 x 3 = 63 or 8 x 3 = 24	B 8 x 7 = 56 or 31 x 22 = 682
B 7 x 9 = 63 or 61 x 12 = 732	B 7 x 7 = 49 or 37 x 11 = 407
B 32 x 42 = 1344 or 4 x 6 = 24	B 54 x 9 = 486 or 8 x 8 = 64
B 22 x 4 = 88 or 6 x 8 = 48	B 51 x 2 = 102 or 8 x 4 = 32

Times 9 Chart

9	x	**1**	=	9
9	x	**2**	=	18
9	x	**3**	=	27
9	x	**4**	=	36
9	x	**5**	=	45
9	x	**6**	=	54
9	x	**7**	=	63
9	x	**8**	=	72
9	x	**9**	=	81

2s and 3s Multiplication Fact Cards

2 x 7 =	14	2 x 8 =
16	2 x 9 =	18
2 x 10 =	20	2 x 1 =
2	2 x 0 =	0
2 x 2 =	4	2 x 3 =

2s and 3s Multiplication Fact Cards

2 x 4 =	8	2 x 5 =
10	2 x 6 =	12
3 x 3 =	9	3 x 0 =
0	3 x 1 =	3
3 x 2 =	6	6

2s and 3s Multiplication Fact Cards

3 x 4 =	12	3 x 5 =
15	3 x 6 =	18
3 x 7 =	21	3 x 8 =
24	3 x 9 =	27
3 x 10 =	30	

4s and 5s Multiplication Fact Cards

4 x 1 =	4	4 x 2 =
8	4 x 3 =	12
4 x 4 =	16	4 x 5 =
20	4 x 6 =	24
4 x 7 =	28	4 x 8 =

4s and 5s Multiplication Fact Cards

32	4 x 9	36
4 x 10	40	4 x 0
0	5 x 1	5
5 x 2	10	5 x 3
15	5 x 4	20

4s and 5s Multiplication Fact Cards

5 x 5 =	25	5 x 6 =
30	5 x 7 =	35
5 x 8 =	40	5 x 9 =
45	5 x 10 =	50
5 x 0 =	0	

Times Cover Game

1	2	3
4	5	6
8	9	10
12	15	16
18	20	24
25	30	36

Multiplication Tables—6s to 9s

×	1	2	3	4	5	6	7	8	9	10
1						6	7	8	9	10
2						12	14	16	18	20
3						18	21	24	27	30
4						24	28	32	36	40
5						30	35	40	45	50
6						36	42	48	54	60
7						42	49	56	63	70
8						48	56	64	72	80
9						54	63	72	81	90
10						60	70	80	90	100

Multiplication by One Digit

1) $\begin{array}{r} 769 \\ \times\ 6 \\ \hline \end{array}$

2) $\begin{array}{r} 742 \\ \times\ 9 \\ \hline \end{array}$

3) $\begin{array}{r} 40 \\ \times\ 4 \\ \hline \end{array}$

4) $\begin{array}{r} 4 \\ \times 5 \\ \hline \end{array}$

5) $\begin{array}{r} 83 \\ \times\ 0 \\ \hline \end{array}$

6) $\begin{array}{r} 643 \\ \times\ 0 \\ \hline \end{array}$

7) $\begin{array}{r} 75 \\ \times\ 8 \\ \hline \end{array}$

8) $\begin{array}{r} 11 \\ \times 2 \\ \hline \end{array}$

9) $\begin{array}{r} 874 \\ \times\ 0 \\ \hline \end{array}$

10) $\begin{array}{r} 24 \\ \times\ 6 \\ \hline \end{array}$

11) $\begin{array}{r} 19 \\ \times\ 0 \\ \hline \end{array}$

12) $\begin{array}{r} 26 \\ \times\ 0 \\ \hline \end{array}$

13) $\begin{array}{r} 65 \\ \times\ 4 \\ \hline \end{array}$

14) $\begin{array}{r} 651 \\ \times\ 7 \\ \hline \end{array}$

15) $\begin{array}{r} 31 \\ \times 7 \\ \hline \end{array}$

16) $\begin{array}{r} 92 \\ \times 7 \\ \hline \end{array}$

Multiplication Facts Checkers

	5×8=		2×5=		8×3=		6×4=
4×5=		3×5=		7×3=		2×3=	
	6×6=		2×2=		5×1=		8×3=
4×4=		7×7=		5×6=		2×7=	
	9×7=		4×2=		3×8=		2×8=
6×5=		3×3=		5×2=		7×5=	
	8×7=		2×3=		3×5=		2×6=
7×6=		8×4=		2×3=		7×3=	

Blank Checkerboard

Word Problem Form (All Operations)

1) READ and understand the problem. (Write needed numbers with their labels; e.g., 4 desks) **WRITE THE ANSWER SENTENCE** with a blank. (e.g., Room has _____ desks.)	**2) DRAW, use number line, or act out problem.**
3) FIND the operation. Will the answer be larger than any of the numbers in the problem? Are the numbers in the problem the same or equal?	**4) CIRCLE the operation.** **Equal Groups** ↓ Larger → ADDITION MULTIPLICATION Smaller → SUBTRACTION DIVISION
5) WRITE THE NUMBER SENTENCE. _____ ☐ _____ = _____	**6) WRITE ANSWER in the blank of the sentence in square 1.**

Addition & Multiplication Story Problems

Use the word problem form in Appendix B to solve the following problems. The important point will to determine whether the numbers in the problem are the same or equal. If so, multiplication will be the most efficient process. Students may solve the problems by repeated addition, however.

1. Gale promised to bring fruit to the picnic. He brought 3 oranges, 7 apples, 9 pears and 5 peaches. How many pieces of fruit did he bring?

   Gale brought _____ pieces of fruit to the picnic.

2. Marquita bought 3 DVD's each week for 7 weeks. How many DVD's did she buy?

   Marquita bought _____ DVD's in 7 weeks.

3. Nancy made 4 potato pancakes for Greg, 4 pancakes for Cindy, 4 pancakes for Linda, 4 pancakes for Barbara and 4 pancakes for Gale. How many potato pancakes did she make?

   Nancy made _____ pancakes.

4. Leslie collected toy pigs. She put 3 pigs in each of the 5 windows of her house. How many pigs does she have in the windows of her house?

   Leslie has ____ pigs in the windows of her house.

5. Jennie had 2 gerbils. Those gerbils had 6 baby gerbils in October; in December they had 5 more baby gerbils. How many gerbils does Jennie have now?

   Jennie has _____ gerbils now.

6. Ann took 3 pictures of her family every day of their vacation. They were on vacation for 8 days. How many pictures did Ann take while they were on vacation

   Ann took _____ pictures on her vacation.

7. Tom has 2 pair of boots, 3 regular pairs of shoes and 2 pair of gym shoes. How many pairs of shoes does he have

   Tom has ____ many pair of shoes.

8. Using the numbers in problem 7, how many actual shoes does Tom have? Remember that a pair of shoes has 2 shoes in it.

   Tom has _____ actual shoes.

9. Ellen went to the library 2 times a week. She got 4 books every time that she went. How many books did she get in a week?

   Ellen got _____ books from the library in a week.

10. Kathy gave two pencils to each of her 7 friends. How many pencils did she give away altogether?

    Kathy gave away _____ pencils.

Simple Calculator Division Problems

1) $5\overline{)60}$

2) $4\overline{)40}$

3) $2\overline{)44}$

4) $8\overline{)72}$

5) $3\overline{)12}$

6) $2\overline{)10}$

7) $1\overline{)64}$

8) $6\overline{)54}$

9) $7\overline{)63}$

10) $2\overline{)24}$

11) $6\overline{)90}$

12) $2\overline{)28}$

Division Tic-Tac-Toe Game

Game 1

9 ÷ 3	8 ÷ 2	12 ÷ 4
21 ÷ 3	15 ÷ 3	3 ÷ 1
18 ÷ 3	30 ÷ 6	6 ÷ 3

Game 2

15 ÷ 3	10 ÷ 2	12 ÷ 4
20 ÷ 5	6 ÷ 3	7 ÷ 1
14 ÷ 2	30 ÷ 5	16 ÷ 2

Division Tic-Tac-Toe Game

Game 3

18 ÷ 2	4 ÷ 2	25 ÷ 5
24 ÷ 8	15 ÷ 5	10 ÷ 1
8 ÷ 2	30 ÷ 5	12 ÷ 4

Game 4

21 ÷ 7	14 ÷ 2	16 ÷ 4
24 ÷ 3	15 ÷ 5	7 ÷ 1
24 ÷ 8	18 ÷ 6	27 ÷ 9

Division Tic-Tac-Toe Game

Calculator Division Problems

1)

$6\overline{)78}$

2)

$5\overline{)305}$

3)

$9\overline{)369}$

4)

$7\overline{)630}$

5)

$2\overline{)66}$

6)

$1\overline{)498}$

7)

$2\overline{)86}$

8)

$8\overline{)480}$

9)

$5\overline{)130}$

10)

$2\overline{)622}$

11)

$2\overline{)74}$

12)

$1\overline{)225}$

Division Problems Checked by Multiplication

1)

$$4\overline{)16}$$

2)

$$7\overline{)84}$$

3)

$$3\overline{)24}$$

4)

$$4\overline{)28}$$

5)

$$5\overline{)20}$$

6)

$$8\overline{)96}$$

7)

$$3\overline{)12}$$

8)

$$6\overline{)12}$$

9)

$$3\overline{)99}$$

10)

$$8\overline{)48}$$

11)

$$5\overline{)25}$$

12)

$$9\overline{)54}$$

13)

$$7\overline{)84}$$

14)

$$3\overline{)21}$$

15)

$$9\overline{)63}$$

16)

$$8\overline{)16}$$

Fraction Strips

Pieces	
1	
2	
4	
8	
16	
10	
5	
3	
6	
12	

Domino Fraction Board—1

$\overline{6}$s	$\overline{2}$s & $\overline{3}$s	$\overline{4}$s	$\overline{5}$s
$\dfrac{5}{6}$	$\dfrac{1}{3}$	$\dfrac{1}{4}$	$\dfrac{1}{5}$
$\dfrac{2}{6}$	$\dfrac{2}{3}$	1	$\dfrac{2}{5}$
$\dfrac{3}{6}$	$\dfrac{1}{2}$	$\dfrac{3}{4}$	$\dfrac{3}{5}$
$\dfrac{4}{6}$	1	$\dfrac{2}{4}$	$\dfrac{4}{5}$

Domino Fraction Board—2

$\overline{6}$s	$\overline{2}$s & $\overline{3}$s	$\overline{4}$s	$\overline{5}$s
$\dfrac{3}{6}$	$\dfrac{2}{3}$	1	$\dfrac{4}{5}$
$\dfrac{1}{6}$	$\dfrac{1}{3}$	$\dfrac{1}{4}$	1
$\dfrac{2}{6}$	1	$\dfrac{2}{4}$	$\dfrac{2}{5}$
$\dfrac{5}{6}$	$\dfrac{1}{2}$	$\dfrac{1}{4}$	$\dfrac{3}{5}$

Domino Fraction Board—3

$\overline{6}$s	$\overline{2}$s & $\overline{3}$s	$\overline{4}$s	$\overline{5}$s
$\dfrac{4}{6}$	$\dfrac{2}{3}$	$\dfrac{2}{4}$	$\dfrac{3}{5}$
$\dfrac{5}{6}$	1	$\dfrac{1}{4}$	$\dfrac{1}{5}$
$\dfrac{3}{6}$	$\dfrac{1}{2}$	$\dfrac{3}{4}$	$\dfrac{2}{5}$
$\dfrac{2}{6}$	$\dfrac{1}{3}$	1	$\dfrac{4}{5}$

Domino Fraction Cards

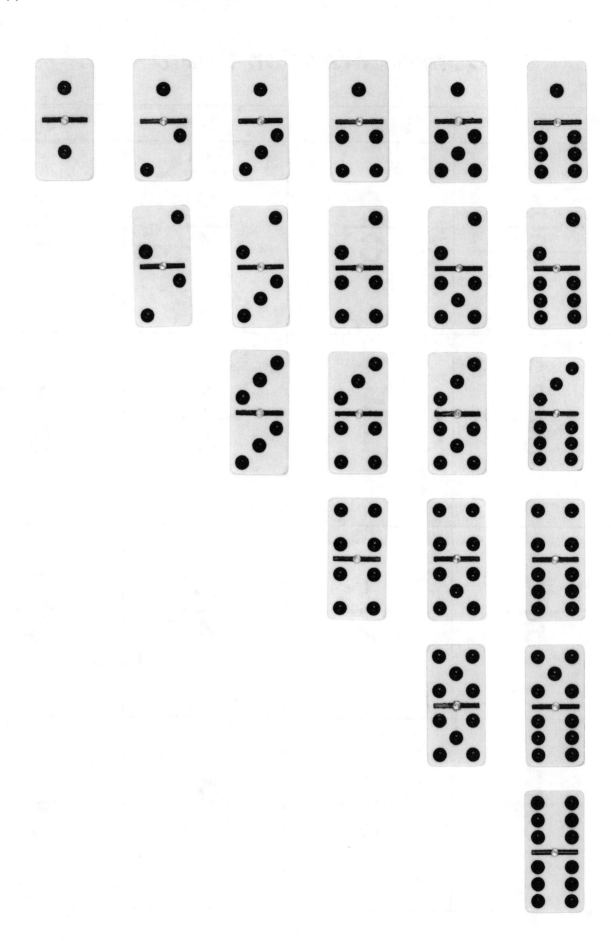

Domino Fraction Cards

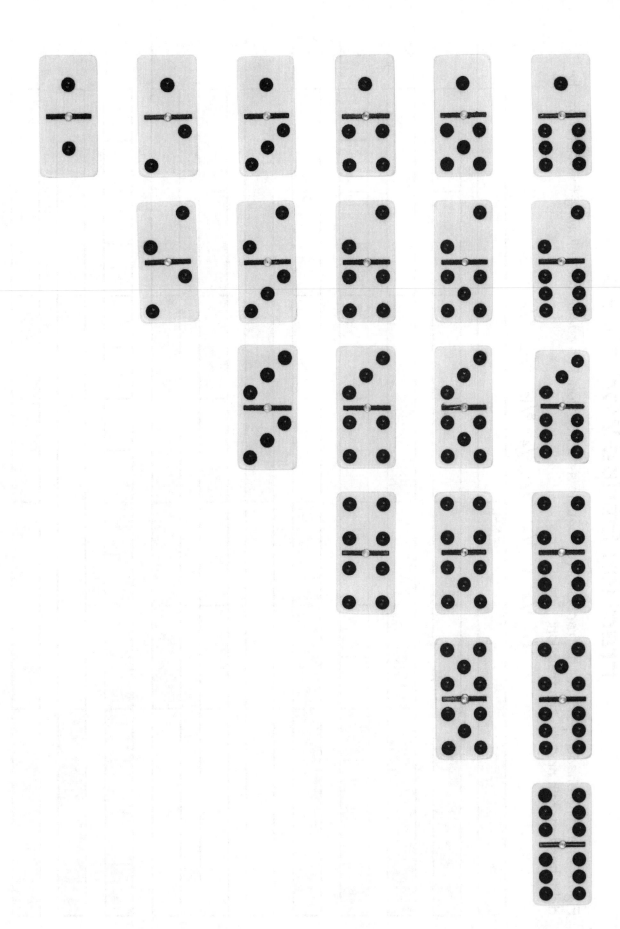

Fraction Strips Worksheet

Figure what the shaded part of the fraction strip equals as a fraction. Put one of the following fractions on the right hand margin next to the fraction:

$1/2$ $1/4$ 1 $3/4$ $6/8$ $2/4$ $8/16$ $4/16$ $4/8$

Equivalent Fractions Worksheet

Use the Fraction Strips Worksheet (from Appendix B) to work these problems.

1)
$$\frac{6}{8} = \frac{}{4}$$

2)
$$\frac{2}{10} = \frac{}{5}$$

3)
$$\frac{4}{6} = \frac{}{3}$$

4)
$$\frac{3}{6} = \frac{}{10}$$

5)
$$\frac{2}{4} = \frac{}{6}$$

6)
$$\frac{4}{6} = \frac{}{9}$$

7)
$$\frac{4}{8} = \frac{}{2}$$

8)
$$\frac{3}{4} = \frac{}{8}$$

9)
$$\frac{1}{5} = \frac{}{10}$$

10)
$$\frac{1}{4} = \frac{}{8}$$

11)
$$\frac{4}{5} = \frac{}{10}$$

12)
$$\frac{2}{5} = \frac{}{10}$$

Perimeter Worksheet

Figure out the perimeter (amount of fence needed) for each rectangle.

1.
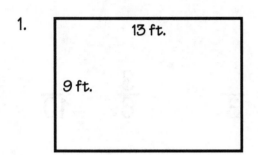

13 ft.

9 ft.

2.
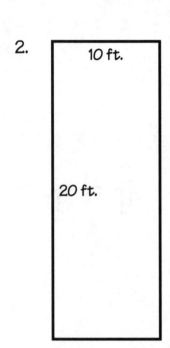

10 ft.

20 ft.

3.
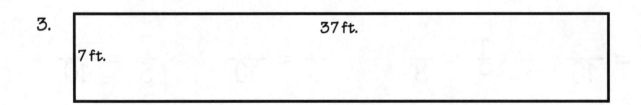

37 ft.

7 ft.

4.

40 ft.

6 ft.

Andy's Play Area

Square Inch Pieces

Perimeter and Area Worksheet

Figure out the **perimeters** (fences) of the following backyards. Then figure the **area** (amount of grass sod needed), in square feet, of the backyards.

EXAMPLE

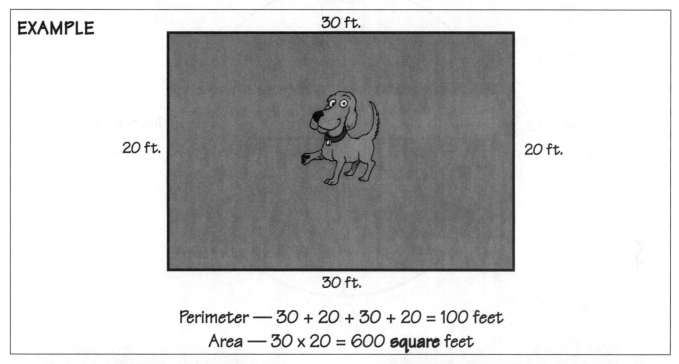

30 ft.

20 ft.

20 ft.

30 ft.

Perimeter — 30 + 20 + 30 + 20 = 100 feet
Area — 30 x 20 = 600 **square** feet

1)

22 ft.

15 ft.

P. = _____   A. = _____

2)

25 ft.

9 ft.

P. = _____   A. = _____

3)

22 ft.

11 ft.

P. = _____   A. = _____

4)

24 ft.

16 ft.

P. = _____   A. = _____

Quarter Clock

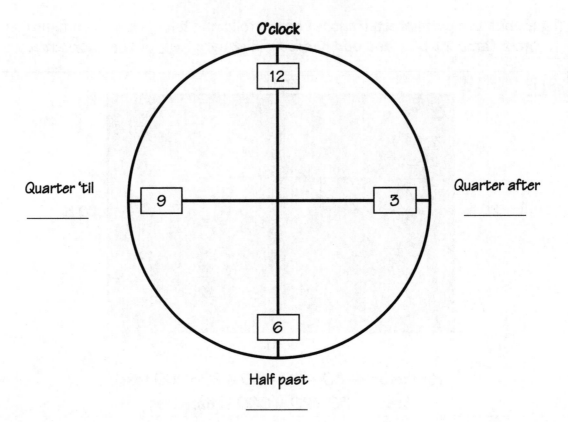

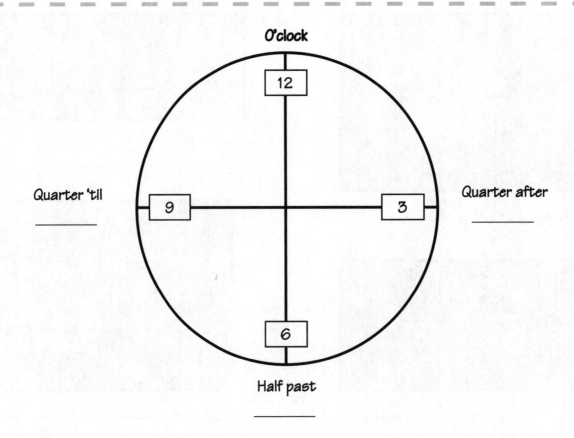

Activity Strips for Time Matching Game

Cut out the activity strips. Put the activity strips in a pile, face down. The player draws one activity strip. He tries to find an appropriate time from the time cards, placed face up on the table. If he is correct, he takes strip and time card. If incorrect, he returns the strip to the bottom of the pile. Another player takes a turn. The winner is the one who has the most activity/time card pairs.

1. Sleep at night.
2. Go to an amusement park and go on rides.
3. Brush your teeth.
4. Dress for school or work.
5. Dress for winter outside (coat, hat, boots, etc.)
6. Make a jelly sandwich.
7. Eat breakfast.
8. Go to and see a movie.
9. Length of one cartoon show on TV.
10. Make your bed.

Activity Strips for Time Matching Game

Game Cards for Time Matching Game

Cut out each time card. Put the time cards, face up on the table.

8+ hours
8+ hours
5 minutes
30 minutes
10-15 minutes
5 minutes
10-15 minutes
2-2½ hours
30-60 minutes
10-15 minutes

8+ hours
8+ hours
5 minutes
30 minutes
10-15 minutes
5 minutes
10-15 minutes
2-2½ hours
30-60 minutes
10-15 minutes

Answer Sheet for Time Matching Game

ACTIVITY	TIME
1. Sleep at night.	8+ hours
2. Go to an amusement park and go on rides.	8+ hours
3. Brush your teeth.	5 minutes
4. Dress for school or work.	30 minutes
5. Dress for winter outside (coat, hat, boots, etc.)	10-15 minutes
6. Make a jelly sandwich.	5 minutes
7. Eat breakfast.	10-15 minutes
8. Go to and see a movie.	2-2½ hours
9. Length of one cartoon show on TV.	30-60 minutes
10. Make your bed.	10-15 minutes

Food Temperature Danger Zone Chart

Safely cold	Food Temperature Danger Zone	Safely Hot
39°F and below (4°C)	40°F.......................140°F RING No more than 2 hours	141°F and above (60°C)

Keep the temperature of hot food above 140°F / 60°C.

Keep the temperature of cold food below 40°F / 4°C.

Do not let food sit at in-between temperatures (between 40 and 140°F or between 4 and 60°C) for more than 2 hours.

Place Value Money Chart

HUNDREDS	TENS	ONES	DECIMAL POINT WALL	CENTS	

Reading Prices Worksheet

If the student can write out the numbers, have them write the numbers as done in example. If not, have the student read the prices out loud to an adult.

HUNDREDS	TENS	ONES	DECIMAL POINT WALL	CENTS	
	read	together		read together	
4	2	3		5	5
Four Hundred	Twenty	Three	AND	Fifty	Five
5	3	2		1	1
2	2	4		7	2
1	0	5		2	2
6	2	2		0	9
4	3	7		4	5
3	5	6		0	3

Blank Checks

Name
Address
Phone

Check Number _____

PAY TO THE
ORDER OF _____

Example $ 4	3	•	1	7
$				

FOR _____

⑆000000000⑆ 00000000⑈

Signature _____

Name
Address
Phone

Check Number _____

PAY TO THE
ORDER OF _____

Example $ 4	3	•	1	7
$				

FOR _____

⑆000000000⑆ 00000000⑈

Signature _____

Name
Address
Phone

Check Number _____

PAY TO THE
ORDER OF _____

Example $ 4	3	•	1	7
$				

FOR _____

⑆000000000⑆ 00000000⑈

Signature _____

Blank Checks

Name
Address
Phone

Check Number _____

PAY TO THE
ORDER OF _____

Example $ 4	3	•	1	7
$				

FOR _____

⑆00000000⑆ 00000000⑈ Signature _____

Name
Address
Phone

Check Number _____

PAY TO THE
ORDER OF _____

Example $ 4	3	•	1	7
$				

FOR _____

⑆00000000⑆ 00000000⑈ Signature _____

Name
Address
Phone

Check Number _____

PAY TO THE
ORDER OF _____

Example $ 4	3	•	1	7
$				

FOR _____

⑆00000000⑆ 00000000⑈ Signature _____

Calculator Money Worksheet

1)
$$\$152.62$$
$$- \$8.06$$

2)
$$\$41.64$$
$$+ \$62.51$$

3)
$$\$184.47$$
$$- \$36.97$$

4)
$$\$7.74$$
$$- \$4.78$$

5)
$$\$201.83$$
$$+ \$116.91$$

6)
$$\$317.11$$
$$- \$81.82$$

7)
$$\$322.96$$
$$+ \$280.60$$

8)
$$\$70.21$$
$$- \$61.63$$

9)
$$\$219.75$$
$$+ \$147.14$$

10)
$$\$111.34$$
$$- \$31.92$$

11)
$$\$152.62$$
$$- \$8.06$$

12)
$$\$274.71$$
$$+ \$128.89$$

Next-Highest-Dollar Worksheet

Sample Items

	NEXT HIGHEST DOLLAR		NEXT HIGHEST DOLLAR
Winter Jacket—$68.66	$69.00	CD—$16.99	
Radio Boom Box—$56.82		TV series on DVD—$35.35	
Gym Shoes—$47.22		Bicycle—$259.69	
Model Railroad Cars—$29.56		(Put in student's preferred item.)	
Jeans—$42.99		(Put in student's preferred item.)	
Backpack—$14.02		(Put in student's preferred item.)	
Shirt—$22.56		(Put in student's preferred item.)	
Tennis Racket—$37.44		(Put in student's preferred item.)	

Large Coin Pictures

Small Coin Pictures

1

5

5, 10

5, 10, 15, 20, 25

Nickels Worksheet

The total amount of money is _____.

Dimes Worksheet

The total amount of money is _____.

Dimes and Nickels Worksheet

The total amount of money is _____.

Quarters Worksheet

The total amount of money is _____.

The total amount of money is _____.

The total amount of money is _____.

Mixed Coin Worksheet

The total amount of money is _____.

The total amount of money is _____.

The total amount of money is _____.

The total amount of money is _____.

Mixed Coin (Including Pennies) Worksheet

The total amount of money is _____.

The total amount of money is _____.

The total amount of money is _____.

The total amount of money is _____.

Two Coin Bingo

$0.20	$0.50	$0.30
$0.15	$0.10	$0.20
$0.30	$0.35	$0.50

$0.35	$0.20	$0.15
$0.10	$0.50	$0.30
$0.15	$0.30	$0.10

Three Coin Bingo

$0.15	$0.25	$0.40	$0.30
$0.45	$0.30	$0.55	$0.35
$0.75	$0.20	$0.60	$0.15
$0.40	$0.55	$0.25	$0.75

$0.35	$0.75	$0.60	$0.40
$0.55	$0.40	$0.20	$0.75
$0.25	$0.45	$0.30	$0.15
$0.60	$0.35	$0.55	$0.20

Totaling Your Shopping List

Total each group of the shopping list. Then total all the groups.

1. 1 loaf of bread $ 1.45
    1 package of cookies $2.69
    1 package of bagels   <u>$ 1.89</u>

2. 1 half gallon of milk $ 1.34
    1 quart of orange juice $ 1.11
    1 package of cheese slices   <u>$2.45</u>

3. 1 can of peaches $ .89
    1 can of pears $1.44
    1 bottle ketchup   <u>$ 1.99</u>

4. 1 head of lettuce $ .77
    10 apples $3.22
    1 bag of green grapes   <u>$ 1.99</u>

Total for List

1. _____

2. _____

3. _____

4. _____

TOTAL: _____

Shopping for One Worksheet

We will assume that the grocery store will give the divided price for one item.
Divide the dollar amount by the number of items to find the price for one item.

1. 2 for $5.00   1 = _____

2. 3 for $7.00   1 = _____

3. 5 for $6.00   1 = _____

4. 8 for $3.00   1 = _____

5. 5 for $5.00   1 = _____

6. 10 for $2.00   1 = _____

7. 6 for $12.00   1 = _____

8. 2 for $4.00   1 = _____

9. 9 for $10.00   1 = _____

10. 3 for $ .60   1 = _____

Start

Which

Price is Best?

Game

Pay
Finish

Cut out **"Which Price is Best?"** cards
and use them for the game.

Which Price is Best? Game Cards

Which price is best?   2 pt.

Store B—8 oz. for $0.24

Store A—20 oz. for $0.40

Popcorn

Which price is best?   2 pt.

Store B—$2.00

Store A—$2.20

Sundae

Which price is best?   2 pt.

Store B—10 oz. for $1.00

Store A—4 oz. for $1.00

Brownies

Which price is best?   2 pt.

Store A—$1.50

Store B—$1.00

Ice Cream Cone

Which price is best?   2 pt.

Store A—2 lb. for $1.00

Store B—4 lb. for $1.60

Peanuts

Which price is best?   2 pt.

Store A—12 oz. for $2.40

Store B—10 oz. for $3.00

Pizza

Which price is best?   3 pt.

Store A—1 lb. for $3.00

Store B—2 lb. for $2.00

Candy

Which price is best?   3 pt.

Store A—2 lb. for $1.00

Store B—3 lb. for $2.40

Apples

Which Price is Best? Game Cards

Which price is best?   2 pt.

Store B—4 lb. for $4.00

Store A—8 lb. for $6.40

Bananas

Which price is best?   2 pt.

Store B—2 lb. for $1.00

Store A—3 lb. for $1.00

Tomatoes

Which price is best?   2 pt.

Store B—10 oz. for $1.00

Store A—12 oz. for $2.40

Lemons

Which price is best?   2 pt.

Store A—2 for $1.00

Store B—2 for $2.00

Doughnuts

Which price is best?   2 pt.

Store A—4 lb. for $8.00

Store B—5 lb. for $15.00

Ham

Which price is best?   2 pt.

Store A—2 lb. for $2.00

Store B—3 lb. for $6.00

Hot Dogs

Which price is best?   3 pt.

Store A—3 lb. for $1.50

Store B—2 lb. for $1.00

Carrots

Which price is best?   3 pt.

Store A—4 oz. for $1.00

Store B—12 oz. for $3.60

Red Hot Peppers

Which Price is Best? Game Card Answers

Which price is best?   2 pt.

Store B—8 oz. for $0.24

Store A—20 oz. for $0.40

Popcorn

Which price is best?   2 pt.

Store B—$2.00

Store A—$2.20

Sundae

Which price is best?   2 pt.

Store B—10 oz. for $1.00

Store A—4 oz. for $1.00

Brownies

Which price is best?   2 pt.

Store A—$1.50

Store B—$1.00

Ice Cream Cone

Which price is best?   2 pt.

Store A—2 lb. for $1.00

Store B—4 lb. for $1.60

Peanuts

Which price is best?   2 pt.

Store A—12 oz. for $2.40

Store B—10 oz. for $3.00

Pizza

Which price is best?   3 pt.

Store A—1 lb. for $3.00

Store B—2 lb. for $2.00

Candy

Which price is best?   3 pt.

Store A—2 lb. for $1.00

Store B—3 lb. for $2.40

Apples

Which Price is Best? Game Card Answers

Which price is best? 2 pt.

Store B—4 lb. for $4.00

Store A—8 lb. for $6.40

Bananas

Which price is best? 2 pt.

Store B—2 lb. for $1.00

Store A—3 lb. for $1.00

Tomatoes

Which price is best? 2 pt.

Store B—10 oz. for $1.00

Store A—12 oz. for $2.40

Lemons

Which price is best? 2 pt.

Store A—2 for $1.00

Store B—2 for $2.00

Doughnuts

Which price is best? 2 pt.

Store A—4 lb. for $8.00

Store B—5 lb. for $15.00

Ham

Which price is best? 2 pt.

Store A—2 lb. for $2.00

Store B—3 lb. for $6.00

Hot Dogs

Which price is best? 3 pt.

Store A—3 lb. for $1.50

SAME

Store B—2 lb. for $1.00

Carrots

Which price is best? 3 pt.

Store A—4 oz. for $1.00

Store B—12 oz. for $3.60

Red Hot Peppers

Grocery Shopping List

Please work with your parents or your supported living provider to choose 4 food items that you can purchase for your family or roommate group. Choose the items, then ask your parents or provider what they estimate as the price for each item. Write their estimated cost on the worksheet in the column labeled Estimated Cost. Add up the estimated cost of the four items and see how much money you need to bring. Have your parents or provider provide you with enough cash to buy the four items.

Then go to the grocery store with your calculator and worksheet. Get the four items at the store. Each time, write the actual cost of each item on your sheet. Before you checkout, add the exact cost of the items on your calculator. Don't forget to put the decimal point in correctly. Then check out. Save your receipt to compare it with your worksheet. If your items are food, there will probably not be tax. Bring your worksheet and receipt to class next week. (There will be a prize!)

Item that you want to buy	Estimated Cost	Actual Cost
Ex., 1 package of red Jell-O	$ .60	($ .57)
1.		
2.		
3.		
4.		
TOTAL	$_____	$_____

Bookstore

Sweet Shop

Food Court

Let's to

Restrooms

Clothing Store

(Cut and join with other half.)

Sports Store

Playground

Go
the Mall

TOYS

**Shoe
Store**

**Music &
Electronics**

Let's Go to the Mall Game Cards

Football	Sundae	Ping Pong Paddles
$9	$2	$9
Bicycle	Brownies	Book on Wizards
$145	$3	$12
Cookie	Ice Cream Cone	Baseball Glove
$1	$2	$30
Guitar	Radio	Sweat Shirt
$70	$55	$18

Let's Go to the Mall Game Cards

Sandals $41	Flip Flops $5	Goggles $3
Swimsuit $35	Men's Shirt $24	Women's Skirt $15
Hamburger & Onion Rings $8	Action Figures $17	Gym Shoes $35
Salad $5	Hiking Boots $88	CD Holder $16

Let's Go to the Mall Game Cards

Headphones	Rollerblades	Computer
$24	$38	$350
Dictionary	Music CD	Children's Picture Book
$27	$12	$12
Pajamas	Dominoes	Toddler Doll
$20	$4	$18
Stuffed Bear	Go to the restroom. Lose one turn.	Go to the playground. Lose one turn.
$8		

Shopping Tally Sheet

AMOUNT ON DEBIT CARD: $200

Item 1—Subtract from $_____
debit card balance
AMT. LEFT ON DEBIT CARD $_____

Item 2—Subtract from $_____
debit card balance
AMT. LEFT ON DEBIT CARD $_____

Item 3—Subtract from $_____
debit card balance
AMT. LEFT ON DEBIT CARD $_____

Item 4—Subtract from $_____
debit card balance
TOTAL MONEY LEFT $_____

Shopping Tally Sheet

AMOUNT ON DEBIT CARD: $_____

Item 1—Subtract from $_____
debit card balance
AMT. LEFT ON DEBIT CARD $_____

Item 2—Subtract from $_____
debit card balance
AMT. LEFT ON DEBIT CARD $_____

Item 3—Subtract from $_____
debit card balance
AMT. LEFT ON DEBIT CARD $_____

Item 4—Subtract from $_____
debit card balance
TOTAL MONEY LEFT $_____

Shopping Tally Sheet

AMOUNT ON DEBIT CARD: $_____

Item 1—Subtract from $_____
debit card balance
AMT. LEFT ON DEBIT CARD $_____

Item 2—Subtract from $_____
debit card balance
AMT. LEFT ON DEBIT CARD $_____

Item 3—Subtract from $_____
debit card balance
AMT. LEFT ON DEBIT CARD $_____

Item 4—Subtract from $_____
debit card balance
TOTAL MONEY LEFT $_____

Shopping Tally Sheet

AMOUNT ON DEBIT CARD: $_____

Item 1—Subtract from $_____
debit card balance
AMT. LEFT ON DEBIT CARD $_____

Item 2—Subtract from $_____
debit card balance
AMT. LEFT ON DEBIT CARD $_____

Item 3—Subtract from $_____
debit card balance
AMT. LEFT ON DEBIT CARD $_____

Item 4—Subtract from $_____
debit card balance
TOTAL MONEY LEFT $_____

Let's Go to the Mall Store Key

Sweet Shop
- Ice cream cone
- Cookies
- Sundae
- Brownies

Sports Store
- Bicycle
- Baseball Glove
- Football
- Swim Goggles
- Swimsuit
- Ping Pong Paddles
- Rollerblades
- Hiking Boots

Clothing Store
- Sweatshirt
- Men's Shirt
- Pajamas
- Women's Skirt
- Swimsuit

Bookstore
- Book on Wizards
- Children's Picture Book
- Dictionary

Music & Electronics Store
- CD Holder
- Music CD
- Guitar
- Headphones
- Radio
- Computer

Shoe Store
- Gym Shoes
- Hiking Boots
- Flip Flops
- Sandals

Toy Store
- Bicycle
- Dominoes
- Action Figures
- Stuffed Bear
- Toddler Doll

Food Court
- Hamburger & Onion Rings
- Salad

Estimating Sales Tax Worksheet

Estimate the amount of sales tax needed for a purchase. Use 10% because it can be figured quickly by moving the decimal point one place to the left. Most state sales tax will be less than 10%. You may also have the student total the Price and Estimated Tax.

	Price	Estimated Tax	Total (if desired)
1.	$27.30	$2.73	$30.03
2.	$14.53	_____	_____
3.	$77.70	_____	_____
4.	$36.50	_____	_____
5.	$12.75	_____	_____
6.	$28.32	_____	_____
7.	$23.00	_____	_____
	TOTAL:	_____	_____

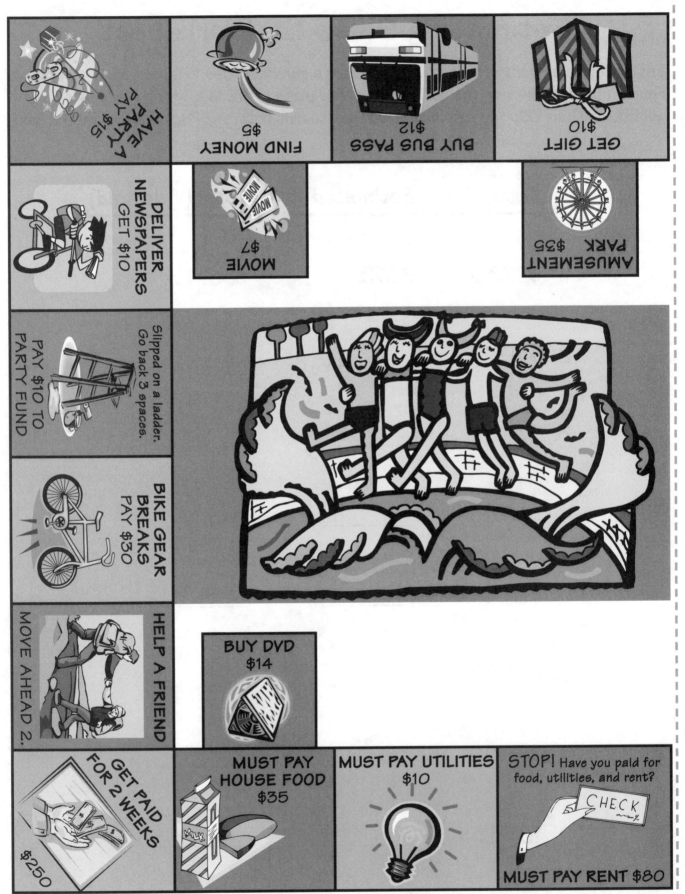

HAVE A PARTY PAY $15

FIND MONEY $5

BUY BUS PASS $12

GET GIFT $10

DELIVER NEWSPAPERS GET $10

MOVIE $7

AMUSEMENT PARK $35

PAY $10 TO PARTY FUND

Slipped on a ladder. Go back 3 spaces.

BIKE GEAR BREAKS PAY $30

HELP A FRIEND

MOVE AHEAD 2.

BUY DVD $14

GET PAID FOR 2 WEEKS $250

MUST PAY HOUSE FOOD $35

MUST PAY UTILITIES $10

STOP! Have you paid for food, utilities, and rent?

CHECK

MUST PAY RENT $80

(Cut and join with other half.)

Where Does the Money Go?

#	Date	What is it?	Pay (-)		Earn (+)		Balance	
			Dollars	Cents	Dollars	Cents	Cents	Cents
TOTAL								

Budget Planning Worksheet (Weekly)

(essentially for items that the student can control)

Planned Expenses: Cost

1._____ $ _____

2. _____ $ _____

3. _____ $ _____

4. _____ $ _____

My Money: Cost

Saved Money $ _____

New Money (job earnings, allowance) $ _____

Other Money (gifts, etc.) $ _____

   TOTAL Money Available to Spend   **$ _____**

What I Actually Spent: (can use initials for items)
        Item   Cost

1._____ $ _____

2. _____ $ _____

3. _____ $ _____

4. _____ $ _____

        TOTAL SPENT   **$ _____**

I had enough money. _____ (yes or no)

Wants or Needs Worksheet

Place the items below in the correct column. Add other items that apply to the students.

- milk
- movie ticket
- bed
- _____

- juice
- macaroni & cheese
- i-Pod
- _____

- potato chips
- underwear
- cable TV
- _____

- rent
- moped
- _____
- _____

NEEDS	WANTS

(There may be variations on what the students think are needs or wants. Have the student explain why that item is a need and you make a judgment on it.)

Banking Deposit Slips

DEPOSIT TO THE ACCOUNT OF:

_____

BANK of _____

09876540:012345678

	DOLLARS	CENTS
CURRENCY		
CHECKS		
LESS CASH		
DEPOSIT TOTAL		

DEPOSIT TO THE ACCOUNT OF:

_____

BANK of _____

09876540:012345678

	DOLLARS	CENTS
CURRENCY		
CHECKS		
LESS CASH		
DEPOSIT TOTAL		

DEPOSIT TO THE ACCOUNT OF:

_____

BANK of _____

09876540:012345678

	DOLLARS	CENTS
CURRENCY		
CHECKS		
LESS CASH		
DEPOSIT TOTAL		

Simulated Payroll Check

Main Street Bakery				27155

po756	Taylor Brown	123-45-6789	02/02/08	02/14/08
Emp. No.	Employee Name	Social Security No.	Period Beginning	Period Ending

Earnings	Hrs	Current Amount	Year to Date	Deductions	Current Amount	Year to Date
$12.00	40	$480.00	$1,440.00	Federal	$4.81	$14.43
				FICA	$29.76	$89.28
				Medicare	$6.96	$20.88
				State	$13.55	$40.65
				City	$4.80	$14.40
				County	$2.40	$7.20
				Health	$71.05	$213.15
				Disability	$2.25	$6.75

$12.00	$480.00	$135.58	$344.42	$1440.00	$406.74	$1033.26
Pay Rate	Current Earnings	Current Deductions	Net Pay	YTD Earnings	YTD Deductions	YTD Net Pay

Main Street Bakery   27155
1212 Main Street
Mainville, USA

PAY   *Three Hundred Fourty-Four and 42/100*

TO THE
ORDER OF   *Taylor Brown*   Date   Amount
                                              02/22/08 $344.42

FIRST BANK OF MAINVILLE
MAINVILLE, USA

||:02155 678 |:072006789|:1236978

Check Forms

DATE_____ 101

PAY TO THE
ORDER OF _____ $ _____
_____ DOLLARS

Bank
1 Main Street
Anywhere, US 10001

FOR _____   _____

⑊17138066⑊   2287057592⑊   0101⑊

DATE_____ 101

PAY TO THE
ORDER OF _____ $ _____
_____ DOLLARS

Bank
1 Main Street
Anywhere, US 10001

FOR _____   _____

⑊17138066⑊   2287057592⑊   0101⑊

DATE_____ 101

PAY TO THE
ORDER OF _____ $ _____
_____ DOLLARS

Bank
1 Main Street
Anywhere, US 10001

FOR _____   _____

⑊17138066⑊   2287057592⑊   0101⑊

Writing Out Check Numbers in Words

Write out in words the money amounts that are given here. Do not forget that the tens and ones are written together as a two digit number. The cents are also written as a two digit number. Here are some of the words that you may have to spell:

- hundred
- fifty
- ninety

- twenty
- sixty
- thousands (later)

- thirty
- seventy

- forty
- eighty

THOUSANDS (Grown-ups)	HUNDREDS (Grown-ups)	TENS (Teens)	ONES (Kids)	DECIMAL POINT WALL	CENTS	
FOR EXAMPLE: Numbers	2	3	5	•	6	6
Written-out Numbers	Two Hundred	Thirty	Five	AND	Sixty	Six
(Write out this number in words on line below.)	4	1	1		3	7
	7	9	1		7	2
	8	9	0		1	0
	3	4	5		0	0

Thousands are the grown-ups that are bigger than hundreds. One thousand is 10 times larger than one hundred. You need to say the grown-up thousands name (One **thousand**) when you read or write out the number.

THOUSANDS (Grown-ups)	HUNDREDS (Grown-ups)	TENS (Teens)	ONES (Kids)	DECIMAL POINT WALL	CENTS	
FOR EXAMPLE: Numbers	2	3	5	•	6	6
Written-out Numbers	Two Hundred	Thirty	Five	AND	Sixty	Six
(Teacher or parent: write in numbers that the student would likely use.)						

Shape Cards

Measuring Angles Worksheet

Note: A ray is a part of a line. Two rays that meet at one point make an angle.

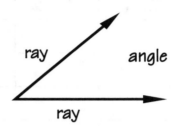

ray   angle

ray

Measure the following angles with a protractor:

1.

2.

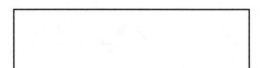

3.

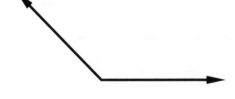

4.

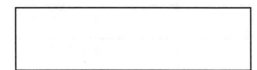

Data Form

_____(title)

	Mon.	Tues.	Wed.	Thurs.	Fri.	Sat.	Sun.
5 hr.							
4½ hr.							
4 hr.							
3½ hr.							
3 hr.							
2½ hr.							
2 hr.							
1½ hr.							
1 hr.							
½ hr.							

_____(title)

	Mon.	Tues.	Wed.	Thurs.	Fri.	Sat.	Sun.
5 hr.							
4½ hr.							
4 hr.							
3½ hr.							
3 hr.							
2½ hr.							
2 hr.							
1½ hr.							
1 hr.							
½ hr.							

Decimal Tenths Strips

Decimal Tenths Strips

Hundredths Chart using pennies

									10
									20
									30
									40
									50
									60
									70
									80
									90
									100

Comparing Decimals

A Hundredths—a whole divided into 100 parts

Shaded on 20 parts out of 100 =

.20

B Tenths—a whole divided into 10 parts (strips)

Shaded on 2 parts out of 10 =

.20

C One Whole—1 unit

Put the shaded parts of the square in order by size:

The square with the biggest shaded part is _____.

The square with the middle-sized amount of shaded area is _____.

The square with the smallest amount of shaded area is _____.

Place Value Fat Mat

TENS	ONES	DECIMAL POINT WALL	TENTHS	HUNDREDTHS
	(read together)			(read together)

Place Value Fat Mat—Money

THOUSANDS	HUNDREDS	TENS (read together)	ONES	DECIMAL POINT WALL	CENTS (read together)	

Percents to Decimals to Worksheet #1

1. 52% = _____

2. 2% = _____

3. 30% = _____

4. 9% = _____

5. 87% = _____

6. 66% = _____

7. 5% = _____

8. 60% = _____

9. 14% = _____

10. 7% = _____

Match 'Em Game Board & Custom Die

20%	33%	25%
10%	1%	5%
33%	10%	20%

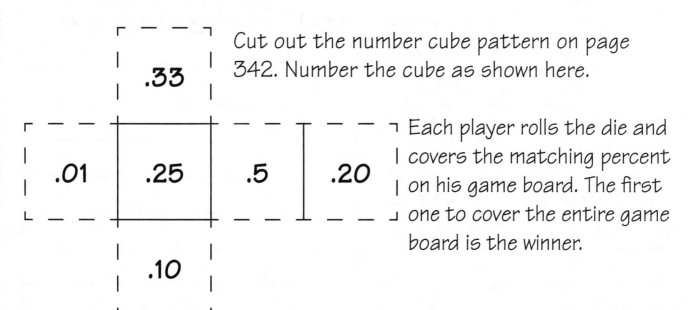

.33

Cut out the number cube pattern on page 342. Number the cube as shown here.

.01 | .25 | .5 | .20

Each player rolls the die and covers the matching percent on his game board. The first one to cover the entire game board is the winner.

.10

Percents to Decimals to Worksheet #2

Remember that the % sign stands for 100ths. There are two ways to change a percent to a decimal:

1. Divide the original number by 100. **Example: 50 ÷ 100 = .50**
2. Move the decimal point two places to the left. (This is the quick way.) Example: Start with 50 and move the decimal point two places to the left until you get to .50.

PART 1: Change the following percents to decimals.

1. 70% = _____	5. 50% = _____	9. 85% = _____
2. 25% = _____	6. 75% = _____	10. 10% = _____
3. 40% = _____	7. 45% = _____	11. 15% = _____
4. 90% = _____	8. 55% = _____	12. 60% = _____

PART 2: Change the following percents to decimals. Then multiply the decimal number times the original price of the item. The answer will be the amount of money you save if you buy the item on sale. (You have to subtract that amount from the original price to get the final amount of money that you pay, but we are not doing that in this worksheet.)

#	Original price	% off (convert percent to decimal)	Amount of money saved (% off times original price)
13.	$125.00	75% = _____	
14.	$40.00	20% = _____	
15.	$80.00	60% = _____	
16.	$100.00	40% = _____	
17.	$60.00	10% = _____	
18.	$16.00	**25% =** _____	
19.	$50.00	55% = _____	
20.	$44.00	65% = _____	

Flap Chart for Percents, Fractions, and Decimals

Top flap

$\frac{1}{2}$	$\frac{1}{4}$	$\frac{3}{4}$	$\frac{1}{3}$	$\frac{2}{3}$	$\frac{1}{5}$	$\frac{1}{6}$

Bottom piece

.5 50%	.25 25%	.75 75%	.33 33%	.67 67%	.20 20%	.17 83%

Include 1/5 and 1/6 only if useful in student's environment

Fold a piece of cardstock in half lengthwise. Staple the top at the fold. Glue the fraction strip above on the top of the "book" and the decimal and percent equivalents inside. Cut the top flap on the lines between the fractions. You should be able to lift up each flap (fractions) and see the equivalent percents and decimals on the second page.

Go Fish—Math Version

$^1/_4$	25%	.25
$^1/_2$	50%	.50
$^3/_4$	75%	.75
$^1/_3$	33%	.33

$2/3$	67%	.67
$1/5$	20%	.2 or .20
$1/6$	17%	.17

Use 1/6 only if students will have use for it.

When first introducing the game, color each set of 3 so they have another cue for learning. Make another set with no color when they have learned more.

Word Problem Cue Sheet

NOT Equal Numbers	Equal Numbers
Answer Larger ADDITION	Answer Larger MULTIPLICATION
Answer Smaller SUBTRACTION	Answer Smaller DIVISION

Word Problems

The student should be able to solve single-step word problems using the structured form. For this exercise, the student will be reminded of the question in the form and will be given the answer sentence with a blank, but he will not use the structured form. If he is unable to solve the problems that follow, make copies of the structured form and have him solve the problems using the forms.

The Hakes family wanted to plan a surprise birthday party for their Aunt Laurie. They planned to invite most of their relatives and many of her friends. They had to plan many things.

1. Laurie's sister, Ann, invited 12 relatives and 11 of Laurie's friends. How many guests were invited?

   ■ Will the answer be larger than the number of relatives or the number of friends?

   ■ Ann invited _____ guests to the party.

2. 10 relatives and 10 friends were able to come to the birthday party. How many guests could come?

   ■ Will the answer be larger than the number of friends or the number of relatives who could come?

   ■ _____ guests could come to the party.

3. There are 5 people in the Hakes family. How many places should Ann set at the dinner table? (Remember that the family and all the guests each need a place at the dinner table.)

   ■ Will the number of places at the table be larger than the number of guests or the number of people in the family?

   ■ Ann will set _____ places at the dinner table.

4. Every member of the Hakes family (5) got 3 presents for Laurie. How many presents did Laurie get from the Hakes family?

   ■ Will the answer be larger than the number of people in the family or the number of presents they each bought?

   ■ Laurie got _____ presents from the Hakes family.

5. Katri bought 3 packages of hamburger buns for the party. Each package had 24 buns in them. How many buns did Katri buy?

   - Will the answer be larger than the largest number in the problem? Are there equal sized groups of numbers?

   - Katri bought _____ hamburger buns

6. Heidi bought 10 pounds of hamburger. If each pound of hamburger makes 4 hamburgers, how many hamburgers would the 10 pounds of hamburger make?

   - Will the answer be larger than the largest number in the problem? Are there equal sized groups of numbers?

   - 10 pounds of hamburger will make _____ hamburgers for the party

7. Heidi wants to make 48 cups of punch for the party. If there are 16 cups in a gallon, how many gallons of punch does she need to make?

   - Will the answer be larger than the largest number in the problem? Are there equal sized groups of numbers?

   - Heidi needs to make _____ gallons of punch for the party.

8. Jeff bought a cake mix for $1.10, frosting for $1.00, birthday candles for $1.00, and ice cream for $4.00. How much did Jeff spend for the party food?

   - Will the answer be larger than the largest number in the problem? Are there equal sized groups of numbers?

   - Jeff paid $_____ for food for the party

9. Katri bought balloons for the party for $3.00. She gave the clerk a 5 dollar bill. How much change should the clerk give her back?

   - Will the answer be larger than the largest number in the problem? Are there equal-sized groups of numbers?

   - The clerk gave Katri $_____ back in change.

10. Laurie was very surprised when she came home and all the people yelled, "Surprise!" There were 26 people at the party. An hour later, Ann counted 20 people still at the party. How many people had left the party by then?

    - Will the answer be larger than the largest number in the problem? Are there equal sized groups of numbers?

    - _____ people had left the party an hour later.

Finding the Average

1. Heinz got 90 on the science test. Kathy got 80, Ruth got 70, Carol got 70, and Gale got a score of 80. What was the average score of the group?

   Rewrite the answer sentence: The average score of the group was _____.
   Add all the scores together: 90 + 80+ 70 + 70 + 80 = **390**
   Divide the answer by the number of scores: 390 divided by 5 = 78
   Put the answer in the answer sentence: The **average score of the group was 78.**

2. The scores on the spelling test were 90, 100, 90, 70, 100, and 90. What was the average score on the spelling test?

   Rewrite the answer sentence:
   Add:
   Divide:
   Put in answer:

3. Allen ate 3 potato pancakes. Scott ate 6, Steve ate 2, Brenda ate 2, and Bryan ate 2 pancakes. What was the average number of pancakes that were eaten?

4. McKenzie sold 18 packages of cookies, Shelby sold 12 packages, Torrie sold 20 packages, Veronica sold 13 packages, Rachel sold 11 packages, and Madison sold 10 packages. What was the average number of packages of cookies that the girls sold?

5. Kristi bought 22 holiday presents, Jennie bought 25 presents, Greg bought 30 presents, Cindy bought 25 presents, Maria bought 20, Barbara bought 28, and Linda bought 4 presents. What was the average number of presents bought?

6. Dana borrowed 3 books from the library, Sean borrowed 4, Sandy borrowed 3 books, Jabari borrowed 5, Rhonda borrowed 5, and Tina borrowed 4 books. What was the average number of books borrowed?

7. On vacation Mark took 33 pictures, Heather took 22, Paul took 40 pictures, Karryl took 24, Scott took 20, Jeff took 45 pictures, and Heidi took 12 pictures. What was the average number of pictures that each took?

ANSWERS

Appendix B8
1—18
2—12
3—13
4—4
5—9
6—16
7—2
8—16
9—11
10—17

Appendix B10
1—7
2—189
3—184
4—106
5—46
6—61
7—107
8—132
9—102
10—81
11—95
12—96

Appendix B14
1—18
2—7
3—442
4—12
5—54
6—14

Appendix B16
1—5
2—4
3—9
4—2

Appendix B17
1—5
2—9
3—23
4—1
5—25
6—17
7—19
8—3
9—21
10—8

Appendix B18
1—HM
2—HC
3—C
4—T
5—HC
6—T

Appendix B19
1—935
2—55
3—14
4—14
5—52
6—18
7—125
8—5
9—673
10—11
11—334
12—5

Appendix B21
1—35
2—13
3—6
4—7
5—11

Appendix B22
1—65
2—6
3—37
4—95
5—33
6—22
7—12
8—26

Appendix B27
1—50
2—30
3—20
4—40
5—60
6—70
7—40
8—80

Appendix B28, page 1
1—0
2—0
3—56
4—190
5—128
6—186
7—61
8—67
9—462
10—423
11—40
12—0

ANSWERS

Appendix B28, page 2

1—0
2—192
3—3920
4—3856
5—378
6—72
7—455
8—7920
9—131
10—4032
11—5598
12—1100

Appendix B36

1—4614
2—6678
3—160
4—20
5—0
6—0
7—600
8—22
9—0
10—144
11—0
12—0
13—260
14—4557
15—217
16—644

Appendix B40

1—24
2—21
3—20
4—15
5—13
6—24
7—7
8—14
9—8
10—14

Appendix B41

1—12
2—10
3—22
4—9
5—4
6—5
7—64
8—9
9—9
10—12
11—15
12—14

Appendix B42

Game 1

3	4	3
7	5	3
6	5	2

Game 2

5	5	3
4	2	7
7	6	8

Game 3

9	2	5
4	3	10
4	6	6

Game 4

3	7	4
8	3	7
3	3	3

ANSWERS

Appendix B43
1—13
2—61
3—41
4—90
5—33
6—498
7—43
8—60
9—26
10—311
11—38
12—225

Appendix B44
1—4
2—12
3—8
4—7
5—4
6—12
7—4
8—2
9—33
10—6
11—5
12—6
13—12
14—7
15—7
16—2

Appendix B49
1—3
2—1
3—2
4—5
5—3
6—6
7—1
8—6
9—2
10—2
11—8
12—4

Appendix B50
1—44
2—60
3—88
4—92

Appendix B53
1) P=74   A=330
2) P=68   A=225
3) P=66   A=242
4) P=80   A=384

Appendix B62
1—$144.56
2—$104.15
3—$147.50
4—$2.96
5—$318.74
6—$235.29
7—$603.56
8—$8.58
9—$366.89
10—$79.42
11—$153.56
12—$403.60

Appendix B63

69	17
57	36
48	260
30	
43	
15	
23	
38	

Appendix B66
30

Appendix B67
60

Appendix B68
45

Appendix B69
50
100
75

Appendix B70
40
60
80
90

Appendix B71
61
55
78
72

ANSWERS

Appendix B73

1—$6.03
2—$4.90
3—$4.32
4—$5.98
5—$21.23

Appendix B74

1—$2.50
2—$2.34
3—$1.20
4—$0.38
5—$1.00
6—$0.20
7—$2.00
8—$2.00
9—$1.11
10—$0.20

Appendix B93

1—90
2—45
3—135
4—30

Appendix B63

13	.75	93.75
14	.20	8.00
15	.60	48.00
16	.40	40.00
17	.10	6.00
18	.25	4.00
19	.55	27.50
20	.65	28.60

Appendix B106

1—yes, 23 guests
2—yes, 20 guests
3—yes, 25 plates
4—yes, 15 presents
5—yes, yes, 72 buns
6—yes, yes,
    40 hamburgers
7—no, yes
    3 gallons
8—yes, no, $7.10
9—no, no, $2.00
10—no, no
    6 people

Appendix B107

2—90
3—3
4—14
5—22
6—4
7—28

Appendix C: Baking with Various Fractions

BANANA BREAD

INGREDIENTS:
- ⅓ cup of butter or margarine
- ⅔ cup of sugar
- 2 eggs beaten
- 1 teaspoon vanilla
- 2 tablespoons milk
- ½ teaspoon salt
- 1⅔ cups flour
- 1 teaspoon baking soda
- 1 teaspoon baking powder
- 1 cup ripe bananas, crushed
- ½ cup nuts, if desired

DIRECTIONS:
1. Mix butter and sugar until smooth in a large bowl.
2. Add 2 beaten eggs and mix.
3. Add ½ teaspoon salt, 2 tablespoons of milk, and 1 teaspoon vanilla.
4. In another bowl mix dry ingredients:
   a. 1⅔ cups of flour
   b. 1 teaspoon baking soda
   c. 1 teaspoon baking powder
5. Add dry ingredients to butter, sugar, and egg mixture.
6. Crush bananas with a fork until smooth.
7. Pour banana mixture into flour and sugar mixture and mix.
8. Add nuts (by hand) to mixed bread dough.
9. Grease one loaf pan.
10. Cook at 350 degrees for 50-60 minutes.

APPLESAUCE MUFFINS

INGREDIENTS:

- 1 cup applesauce
- ¾ cup brown sugar
- ½ cup oil
- 1¾ flour
- 1 teaspoon baking soda
- ½ teaspoon salt
- 1 teaspoon cinnamon
- ½ teaspoon cloves
- 1/3 cup nuts
- ½ cup raisins (optional)

DIRECTIONS:

1. Combine in a large bowl and mix well:
   - a. 1 cup applesauce
   - b. ¾ cup brown sugar
   - c. ½ cup oil
2. Mix together in another bowl:
   - a. 1¾ cup flour
   - b. 1 teaspoon baking soda
   - c. ½ teaspoon salt
   - d. 1 teaspoon cinnamon
   - e. ½ teaspoon cloves
3. Mix the dry ingredients into the applesauce mixture and blend. Use electric mixer, if necessary.
4. Add raisins and nuts (optional) to mixture and blend well by hand.
5. Pour into muffin tins, filling cups ⅔ full.
6. Bake at 350 degrees for 30-35 minutes.

FRUITFUL BREAD

INGREDIENTS:

- 1½ cups flour
- 1½ cups mashed bananas
- 1½ teaspoons baking powder
- ½ teaspoon baking soda
- 3 cups cereal with fruit (e.g., Raisin Bran)
- 8 ounce can crushed pineapple, drained
- ¾ cup sugar
- ½ teaspoon salt
- ⅓ cup oil
- 2 large eggs

DIRECTIONS:

1. In a large bowl, mix:
   - a. 1½ cups flour
   - b. ¾ cup sugar
   - c. ½ teaspoon salt
   - d. 1½ teaspoons baking powder
   - e. ½ teaspoon baking soda
2. In a separate bowl, mix bananas, pineapple, oil, and eggs. Mix well.
3. Add cereal and mix.

4. Stir in dry ingredients, mixing well.
5. Spread in a greased loaf pan.
6. Bake at 350 degrees for 50 minutes.

WAFFLES #2

INGREDIENTS:

- 1¾ cups flour
- 1 tablespoon baking powder
- ½ teaspoon salt
- 1½ tablespoon sugar
- Cooking spray

- 3 eggs
- 1¾ cups milk
- ⅓ cup oil
- ½ teaspoon vanilla

Note: This is a larger recipe than Velvet Waffles (Chapter 11) and uses different fractions.

DIRECTIONS:

1. Heat waffle iron and spray with cooking spray.
2. In large bowl, mix dry ingredients together:
   a. 1¾ cup flour
   b. 1 tablespoon baking powder
   c. ½ teaspoon salt
   d. 1½ tablespoon sugar
3. Crack eggs into small bowl and beat them with a fork.
4. Add oil and milk to the eggs in a small bowl.
5. Mix the liquids into the dry ingredients in the large bowl and stir until smooth.
6. Use a half-cup measure to pour waffle batter onto waffle iron.
7. Cook the waffle, without opening the waffle iron, until no steam is visible.

Makes 6 waffles.

BEST BROWNIES (OVEN)

INGREDIENTS:

- 2 cups sugar
- 1½ teaspoons vanilla
- 4 eggs
- ½ cup melted margarine or butter

- ½ cup cocoa
- 1½ cups flour
- 1 teaspoon salt

DIRECTIONS:

1. Mix together in a large bowl:
   a. 2 cups sugar
   b. ½ cup cocoa
   c. 1½ cup flour
   d. 1 teaspoon salt
2. Melt the margarine and measure ½ cup.
3. Beat the 4 eggs with a fork in a small bowl.
4. Add the 1½ teaspoons vanilla to the eggs.

5. Add the wet ingredients (margarine, eggs, and vanilla) to the dry ingredients. Mix.
6. Add nuts, if desired.
7. Pour into a 9" x 13" baking pan.
8. Bake at 350 degrees for 25-30 minutes. Check the brownies after 20 minutes. Do not over-bake.

STRAWBERRY SMOOTHIE

INGREDIENTS:

- ½ cup milk
- ½ cup plain or vanilla yogurt
- 8 strawberries, hulled, or ½ cup frozen strawberries
- 3 tablespoons sugar (if strawberries are unsweetened)
- 1½ teaspoons vanilla extract
- 6 ice cubes, crushed

DIRECTIONS:

1. Put strawberries, milk, yogurt, sugar, and vanilla extract into a blender. Blend until well combined.
2. Add ice cubes and blend until smooth.
3. Pour into 2 glasses.

TROPICAL SMOOTHIE

INGREDIENTS:

- ¼ cup milk
- 1 cup yogurt (vanilla or fruit flavor)
- 1 orange, peeled and separated, or ½ cup canned mandarin oranges (drained)
- ¼ cup crushed pineapple with juice
- ¼ cup lemon juice

DIRECTIONS:

1. Mix all ingredients together in a blender.
2. Blend until smooth.
3. Pour into 2 glasses.

WHITE GRAPE SMOOTHIE

INGREDIENTS:

- ½ cup white grape juice
- ¼ cup seedless grapes
- ¼ cup crushed pineapple with juice
- ½ cup milk
- ½ cup crushed ice cubes

DIRECTIONS:

1. Mix all ingredients together in a blender.
2. Blend until smooth.
3. Pour into 2 glasses.

ORANGE-BANANA SMOOTHIE

INGREDIENTS:
- ½ frozen banana, cut into chunks
- ½ cup orange or vanilla yogurt
- 3 tablespoons orange juice concentrate
- 1 cup vanilla frozen yogurt
- 1 cup milk
- 1½ cups orange juice
- ½ teaspoon vanilla

DIRECTIONS:
1. Mix all ingredients in a blender.
2. Blend at low setting for one minute.
3. Change to high setting and blend until smooth.
4. Pour into 4 glasses.

BANANA JELL-O SMOOTHIE

INGREDIENTS:
- 1 cup vanilla or plain yogurt
- ½ cup banana, cut into chunks
- 1 small package of Jell-O (gelatin), any flavor
- ½ cup milk
- ½ cup ice cubes, crushed

DIRECTIONS:
1. Mix all ingredients in a blender.
2. Blend until smooth.
3. Pour into 2 or 3 glasses.

References

Benighof, A. M. (1998). SenseAble Strategies: Teaching Diverse Learners. Longmont, CO: Sopris West.

Bird, G. & Buckley, S. (2001). *Number Skills for Individuals with Down Syndrome: An Overview.* Portsmouth, U.K.: The Down Syndrome Educational Trust.

Blenk, K. & Fine, D. L. (1995). *Making School Inclusion Work: A Guide to Everyday Practices.* Cambridge, MA: Brookline Books.

Browder , D.M. & Snell, M.E. (1990). Teaching functional academics. In Snell, M.E. & Brown, F. (Eds.), *Instruction of Students with Severe Disabilities* (5th Ed.), 493-542. Upper Saddle River, NJ: Merrill.

Browder, D.M. & Wilson, B. (2001). *Curriculum and Assessment for Students with Significant Cognitive Disabilities.* New York: Guilford Press.

Ebeling, D. G., Deschenes, C., & Sprague, J. (1994). *Adapting Curriculum and Instruction in Inclusive Classrooms: A Teacher's Desk Reference.* Bloomington: Institute for the Study of Developmental Disabilities.

Fowler, A. E. (1990). Language abilities in Down syndrome. In Cicchetti, D. & Baeghly, M. (Eds.). *Children with Down Syndrome: A Developmental Approach.* New York: Cambridge University Press.

Fowler, A. E. (1995). Linguistic variability in persons with Down syndrome. In Nadel, L. & Rosenthal, D. (Eds.). *Down Syndrome: Living and Learning in the Community.* New York: Wiley-Liss.

Garnett, K. (1998). Math learning disabilities. *Division for Learning Disabilities Journal of CEC* (November).

Gray, C. (1994). *The New Social Story Book.* Arlington, TX: Future Horizons.

Hale, N. (2004). *Keeping a Budget.* (Money Management Series for Teens and Adults with Special Needs). Los Gatos, CA: Special Reads for Special Needs.

Hammeken, P.A. (1995). *Inclusion: 450 Strategies for Success.* Minnetonka, MN: Peytral Publications.

Heddens, J. W. (1990). Improving mathematics teaching by using manipulatives. www.fed.cuhk.edu.hk/~fllee/mathfor/edumath/9706/13hedden.html.

Heddens, J.W. (1986). Bridging the gap between the concrete and the abstract. *Arithmetic Teacher,* 33(6), 14-17.

Hembree, R. & Dessart, D. (1986). Effects of hand-held calculators in pre-college mathematics education: A meta-analysis. *Journal for Research in Mathematics Education,* 17 (2), 83-89.

Horstmeier, D. (2004). *Teaching Math to People with Down Syndrome and Other Hands-on Learners, Book 1.* Bethesda, MD: Woodbine House.

Lowe, M. L. & Cuvo, A. J. (1976). Teaching coin summation to the mentally retarded. *Journal of Applied Behavior Analysis,* 9, 483-489.

National Council of Teachers of Mathematics (2005). *Computation, calculators, and common sense: A position paper.* Available from web site www.nctm.org.

Pallotta, J. (2000). *Reese's Pieces: Count by Fives.* New York: Scholastic.

Pallotta, J. (2001). *The Hershey's Kisses Addition Book.* New York: Scholastic.

Pallotta, J. (2002). *The Hershey's Kisses Subtraction Book.* New York: Scholastic.

Pallotta, J. (1999). *The Hershey's Milk Chocolate Bar Fraction Book.* New York: Scholastic.

Reys, R., Suydam, M. N., & Lindquist, M. M. (1995). *Helping Children Learn Mathematics.* Boston: Allyn and Bacon.

Semple, J. (2007). *Semple Math.* North Attleboro, MA: Semple Math, Inc.

Stainback, S. & Stainback, W. (1992). *Inclusion: A Guide for Educators.* Baltimore: Paul H. Brookes.

Swann, K. (2004). *Solving Word Problems with Pictures.* San Antonio: PCI Education.

Wolpert, G. (1996). The educational challenges inclusion study. Available from the National Down Syndrome Society web site at http://ndss.org/index.php (listed under "Information Topics"—"Education and Schooling.")

Xin, Y.P., Grasso, E., Dipipi-Hoy, C. M., & Jitendra, A. K. (2005). Effects of purchasing skills instruction for individuals with developmental disabilities: A meta-analysis. *Exceptional Children,* 71(4), 379-400.

Resources

Attainment Company
P.O. Box 930160
Verona, WI 53593-0160
800-327-4269
www.attainmentcompany.com
Realistic play money; money-related board games for older students; software for teaching about time and money.

Borenson and Associates
P.O. Box 3328
Allentown, PA 18106
800-993-6284
www.borenson.com
Makers of the *Hands-On Equations* program for teaching algebra.

City Creek Press
P.O. Box 8415
Minneapolis, MN 55408
800-585-6059
www.citycreek.com
Publishers of *Addition the Fun Way* and *Times Tables the Fun Way*.

Creative Publications
Wright Group/McGraw-Hill
12600 Deerfield Pkwy.
Alpharetta, GA 30004
888-205-0444
www.creativepublications.com
Publishers of the *Algebra Lab* program for teaching algebra with manipulatives.

Different Roads to Learning
12 W. 18th St.
New York, NY 10011
800-853-1057; 212-604-9637
www.difflearn.com

Products for helping children with autism learn, including some math games that would be useful for children with Down syndrome and other disabilities, as well as children with autism.

Down Syndrome Educational Trust
The Sarah Duffen Centre
Belmont St.
Southsea, Hampshire
England PO5 1Na
enquiries@downsed.org
www.downsed.org

Numicon materials for teaching numeracy, as well as many monographs on math skills and other issues concerning people with Down syndrome.

EAI Education
118 Bauer Dr.
P.O. Box 7046
Oakland, NJ 07436-7047
www.eaieducation.com

This educational supply company has a wide variety of rulers, counters, calculators, play coins and currency, teaching clocks, educational games, and algebra manipulatives.

ETA/Cuisenaire
500 Greenview Ct.
Vernon Hills, IL 60061
800-875-5985; 847-816-5066
www.etacuisenaire.com

The makers of the Algebra Tiles manipulatives, as well as Cuisenaire rods (color-coded manipulatives for counting, teaching about operations, etc.). Also stocks many varieties of calculators, math games, workbooks, dominos, compasses, etc.

Gander Publishing
412 Higuera St., Suite 200
San Luis Obispo, CA 93401
800-554-1819; 805-541-5523
www.ganderpublishing.com

Publishers of the "On Cloud Nine" math program designed for students with learning disabilities, and also sells special paper designed to help students line numbers up properly when computing and many math workbooks.

Innovative Learning Concepts
6760 Corporate Dr.
Colorado Springs, CO 80919
800-888-9191
www.touchmath.com

Makers of the TouchMath series of teaching products, including manipulatives and workbooks for teaching about addition, subtraction, multiplication, division, fractions, time, and money.

Jackpot Bingo Supplies
800-633-4477
www.jackpotbingosupplies.com
    A source of magnetic bingo chips and wands.

Learning Resources
380 N. Fairway Dr.
Vernon Hills, IL 60061
800-222-3909
www.learningresources.com
    Many good math games, calculators, manipulatives, blank dice, workbooks, and other teaching materials. They also offer products such as an Elapsed Time Line and a Time Equation Solver that may be useful for adults with disabilities who need help with tasks such as figuring out how much time to allow to get ready for an event.

Memory Joggers
888-854-9400
www.memoryjoggers.com
    This company offers programs for teaching the addition and multiplication facts using visual and verbal mnemonics.

Onion Mountain Technology
74 Sextons Hollow Rd.
Canton, CT 06019
860-693-2683
www.onionmountaintech.com
    This company specializes in assistive technology. Offers many varieties of calculators, including talking, the Money Calc, See "N" Solve, and the Coin-u-lator; the Time Timer; clock stamps, number and operation stamps; etc.

PCI Educational Publishing
P.O. Box 34270
San Antonio, TX 78265
800-594-4263
www.pcicatalog.com
    Realistic play money, calculators, Time Dominoes, math board games, software.

Remedia Publications
15887 N. 76th St., Ste. 120
Scottsdale, AZ 85260
800-826-4740
www.rempub.com
    Workbooks designed for special education students with a real-world focus.

Rhymes 'n' Times
P.O. Box 87352
Baton Rouge, LA 70879-8352
888-684-6376
www.rhymesntimes.com
    Offers a multisensory program for teaching the multiplication facts using rhymes and manipulatives.

Semple Math

11 Robert Toner Blvd.

PMB 332, Suite 5

North Attleboro, MA 02763

888-868-6284

www.semplemath.com

Publishers of the Semple Math program, which uses mnemonic devices to help students master basic math concepts.

Silver Lining Multimedia

P.O. Box 544

Peterborough, NH 03458

888-777-0876

www.silverliningmm.com

Offers a "sliding number line" and teaching clocks and timers.

Singing Turtle Press

942 Vuelta del Sur

Santa Fe, NM 87507

505-438-3418

www.singingturtlepress.com

Publishes an algebra survival guide and a prealgebra book called *Pre-Algebra Blastoff* which thoroughly explains the concept of positive and negative numbers using manipulatives. The manipulatives for the negative numbers are circular pieces of foam with a cutout in the middle and the positive numbers are round pieces that fit into the holes in the negative numbers. When you put them (add them) together, they form a round, zero-shaped piece (helping students see that a negative plus a positive equals 0). The method works for subtraction too.

SmallToys.com

www.smalltoys.com

A source for magnetic bingo wands and chips.

Special Reads for Special Needs

14 Stacia St.

Los Gatos, CA 95030

866-553-2042; 408-395-1329 (fax)

www.specialreads.com

Publishers of the *Money Management Series for Teens and Adults with Special Needs*.

Texas Instruments

http://education.ti.com/educationportal

Student calculators by Texas Instruments.

U.S. Bingo

3438 E. Ellsworth Rd.

Ann Arbor, MI 48108

800-254-0773

www.us-bingo.com/bingo-chips.html

A source for magnetic bingo wands and chips.

World Class Learning Materials
406 Main St.
P.O. Box 929
Reisterstown, MD 21136
800-638-6470/ 443-712-0985
www.wclm.com
A manufacturer and publisher of educational materials for students, this company has a variety of math games (including Quizmo), flashcards, timers, manipulatives, and learning kits.

SOFTWARE

Attainment Company
P.O. Box 930160
Verona, WI 53593-0160
800-327-4269
www.attainmentcompany.com
Products include *Match Time* and *Basic Coins* software.

Barnum Software
5191 Morgan Territory Road
Clayton, CA 94517
800-553-9155
www.barnumsoftware.com
Creators of *The Quarter Mile,* good software for drilling math facts at the learner's pace.

Education.com
310-649-8007 (customer service)
877-268-6197 (orders)
www.education.com/knowledgeadventure
Source for Knowledge Adventure software, including the *Math Blaster* series.

EdVenture Software
203-299-0291 (fax)
www.edven.com
Makers of *Gold Medal Math* and *Soccer Math,* software for drilling math facts with a sports twist.

Federal Deposit Insurance Corporation (FDIC)
www.fdic.gov/consumers/consumer/moneysmart
Offers a free CD called *Money Smart: An Adult Education Program.* The CD contains a set of 10 training modules covering basic financial topics such as using a bank, credit, budgeting, the importance of saving, and how interest works. The CD includes both an instructor's guide and a participant's workbook.

IntelliTools, Inc.
1720 Corporate Center
Petaluma, CA 94954
800-899-6687 (US); 800-353-1107 (Canada)
www.intellikeys.com
Programmable adaptable keyboards (IntelliKeys), touch screens; some math software designed for individuals with disabilities focusing on early concepts and solving math problems on the computer.

The Learning Company
c/o Riverdeep
100 Pine St., Suite 1900
San Francisco, CA 94111
800-395-0277
www.learningcompany.com

Producers of the *Mighty Math* software series and *Numbers Up! Volcanic Panic* and *Ultimate Math Invaders* (which provide game-like drill on computation, as well as fractions, decimals, percentage, etc.)

Math Companion

Math Companion for Windows/Mac (by Visions) allows you to custom make math worksheets and written activities involving the four operations, fractions, decimals, money, percents, and word problems. Available from many suppliers, including EAI Education and PCI Educational, listed above.

Teacher Created Resources
6421 Industry Way
Westminster, CA 92683
800-662-4321
www.teachercreated.com

Produces *Math Wizard,* relatively inexpensive software for Mac/PC that allows you to create and print customized math worksheets and answer sheets. Covers addition, subtraction, multiplication, division, currency, time, fractions, and measurement conversions.

MATH GAMES

Many of the games referenced in this book (such as those by Milton Bradley) are easily found in most toy stores. Here is contact information for some companies that carry math games that may not be as widely available in stores. Many of the companies in the section above also carry educational games.

Conceptual Math Media
5728 Lonetree Blvd.
Rocklin, CA 95765
888-433-2810; 916-435-2810
www.conceptualmathmedia.com

Makers of the Equate game (similar to Scrabble, only players form equations rather than words), as well as the Conceptual Bingo series of games (versions cover telling time, decimals, fractions, money, number concepts).

Discovery Toys
www.discoverytoysinc.com

Educational toys and software.

Educational Learning Games
727-786-4850
info@ELGames.com
www.educationallearninggames.com

Thousands of educational games, including a vast array of games for teaching and reinforcing computation, telling time, handling money, using fractions, and solving word problems.

Gamewright

www.gamewright.com

Many of Gamewright's card games involve using math skills in a fun way. For example, Rat a Tat Cat involves trying to collect four cards that have the lowest total possible; Maya Madness involves addition of small positive and negative numbers; Zeus on the Loose involves adding numbers to a total of 100 and recognizing multiples of 10.

International Playthings

75D Lackawanna Ave.

Parsippany, NJ 07054

973-316-5883 (fax)

www.intplay.com

This company makes a number of board games, card games, and puzzles that use early math skills. For example, Leaping Frogs and Bean Bag Toss are both active games that use addition skills.

Jax Ltd.

141 Cheshire Lane

Minneapolis, MN 55441

763-449-9699

www.jaxgames.com

Has a variety of educational games, including Over and Out and Match 'Em (addition games).

Learning Resources

380 N. Fairway Dr.

Vernon Hills, IL 60061

800-222-3909

www.learningresources.com

Learning Resources has games that help children learn a variety of math skills, including addition, subtraction, multiplication, and division; money skills; time telling; fractions.

MindWare

2100 County Rd. C W

Roseville, MN 55113-2501

800-999-0398

www.mindwareonline.com

This company sells educational games and books, including many designed to help with learning addition, subtraction, multiplication, and division facts.

Ravensburger

1 Puzzle Lane

Newton, NH 03858

603-382-3377

www.ravensburger.com/rag/com/presse/usa/index/html

This German toy company has many high quality educational games, many of which focus on math concepts. They include *Number Race* (addition and subtraction) and *Pow Wow* (addition including an unknown number).

Talicor
901 Lincoln Parkway
Plainwell, MI 49080
800-433-GAME; 269-685-2345
www.talicor.com
Offers many educational games, including *MathAnimals* (reinforces addition, subtraction, multiplication), *Moneywise Kids,* and *MathSmart* dominoes.

HELPFUL MATH WEBSITES

Some of these sites require payment to access some or all of the activities available. Again, this is only a small sample of what's available.

Algebra Help
www.algebra-help.info

Funbrain
www.funbrain.com

Improving Education Inc.
www.onlineworksheets.org

LearningPlanet
www.learningplanet.com

Math Fact Café
www.mathfactcafe.com

Math Forum @ Drexel
www.mathforum.com

Math Solutions
www.mathsolutions.com

MathStories.com
www.mathstories.com

Multiplication.com
www.multiplication.com

SchoolExpress
www.schoolexpress.com

Schoolhouse Printables
http://schoolhouseprintable.tripod.com

SuperKids Math Worksheet Creator
www.superkids.com/aweb/tools/math

The Arc of the U.S.
1010 Wayne Av, Ste. 650
Silver Spring, MD 20910
800-433-5255; 301-565-3842
www.thearc.org

Autism Society of America
7910 Woodmont Ave., Ste. 300
Bethesda, MD 20814
800-328-8476
www.autism-society.org

Canadian Down Syndrome Society
811 14th St. NW
Calgary, LA T2N 2A4
www.cdss.ca

Children and Adults with Attention Deficit/Hyperactivity Disorder
8181 Professional Place, Suite 150
Landover, MD 20785
800-233-4050; 301-306-7070
www.chadd.org

Learning Disabilities Association of America
4156 Library Rd.
Pittsburgh, PA 15234
412-341-1515
www.ldanatl.org

Learning Disabilities Association of Canada
323 Chapel St
Ottawa, Ontario K1N 7Z2
613-238-5721; 613-245-5391 (fax)
www.ldac-taac.ca

Learning Disabilities Online
www.ldonline.org

National Council of Teachers of Mathematics
1906 Association Dr.
Reston, VA 20191
703-620-9840
www.nctm.org

National Down Syndrome Congress
1370 Center Dr., Ste. 102
Atlanta, GA 30338
800-232-6372
www.ndsccenter.org

National Down Syndrome Society
666 Broadway
New York, NY 10012
800-221-4602
www.ndss.org

Online Asperger Syndrome Information and Support (O.A.S.I.S.)
www.udel.edu/bkirby/asperger

Positive Behavioral Intervention & Supports Technical Assistance Center
Behavioral Research and Training
5262 University of Oregon
Eugene, OR 97403
541-346-2505
www.pbis.org

Rehabilitation Research & Training Center on Positive Behavioral Support
rrtcpbs@fmhi.usf.edu
http://rrtcpbs.fmhi.usf.edu.statement.htm

Index

About the Author

DeAnna Horstmeier, Ph.D, is an Instructional Resources Consultant at a special education regional resource center in Columbus, Ohio, where she provides educational assistance to parents and educators in teaching strategies and materials for students with special needs. She also presents at and facilitates a series on Cognitive Disabilities and on Autism Spectrum Disorder at the center. She has taught at the Ohio State University in both the areas of special education and speech, language, and communication. Her publications include *Ready, Set, Go–Talk to Me: A Handbook for the Teaching of Prelanguage and Early Language Skills Designed for Parents and Professionals (*with James D. MacDonald*)* and various chapters in other professional publications. She is the mother of a young adult son with Down syndrome, whose needs for independent living skills placed her on the road to finding ways to teach useful math in a hands-on manner.